PRAISE FOR PAUL HEBERT AND

THE SUN ABOVE THE CLOUDS

"I finished reading your book and that is a first for me in about 20 years. I must say, I shed some tears. It had me not leaving it alone until the very last page. You and I have kept in touch since the accident and I have always seen you being very positive and uplifting whenever we met or talked. I really did not understand the level of difficulty you were going through as you always made it sound not bad! Reading this book truly brought home the many challenges you faced then and still face every day. I do not know, or have I ever met two individuals who have inspired me as you and Lorraine do. This book is your life experience and those who read it will truly appreciate the definitions of perseverance, stamina, survival, hope, and family."

—Victor Budzinski
Founder, Former CEO, and now Executive Advisor
Valard Construction Ltd.
Alberta, CA

"Having met and talked to Paul Hebert on a number of occasions I am reminded of an expression I once read: *Experience, Strength, and Hope.* This is what Paul is all about, those three things from a lineman's point of view. If you think that you have big problems, and life is getting you down, you should read Paul's book. He has been there and back in ways that many linemen struggle with, besides surviving a massive 14.4kv. contact."

—Byron Dunn
Editor-in-Chief, *Powerlineman Magazine*

"I first met Paul Hebert at the Huntsman World Senior Games when we were assigned to the same cart for a social golf tournament. We hit it off immediately and have remained friends ever since. Of course, one of the first things I noticed was his physical condition. But that only lasted until he outdrove me on the first hole (and outplayed me the rest of the day). Mostly, though, what I remember are our conversations.

"Paul was doing quite well adjusting to his condition, all things considered. Still, the magnitude of his injuries was a lot to overcome and he was not content to be doing well, *all things considered.* He was interested in doing *well*, period. I admired that. I admired his introspection and his curiosity. It was obvious that Paul had a level of fortitude that helped him cope with injuries that would be devastating to most of us. How he obtained/maintained that fortitude and the challenges he faced/struggled with is a story well worth being told.

"Paul and I have maintained contact with each other since that first meeting and we have talked more about his experiences. I am looking forward to playing more golf with Paul, and, of course, having more great conversations. Paul's journey has been a tough one and I am sure he has more challenges in his future. I am also sure that Paul will face those challenges with the same tenacity that he has had to date.

"I am glad we met, and I am glad he is my friend."

—Thomas A. Burling, Ph.D.
Former Chief, Domiciliary Services
Palo Alto VA Medical Center
Palo Alto, California

"You know, Paul, you made a big difference in my life, helping me in times of need, supporting me, pointing me in the right direction. When I had that girl die in my arms from that car accident, and the surgeries I went through when I got plowed into from behind, you were always there to help pick up broken me, even just in conversation when I needed someone to talk to. And I am proud of all the crews I ran, never to have had an accident with any of them as I taught them and empowered them to succeed. I have a lot of respect for what you helped me achieve. We did it as a team, and for that I thank you! Keep in touch my friend!"

—Brad Kolody

"Knowing you for almost 30 years, you've shared with me in the past a bit of your upbringing and life as a young adult. Putting it together in a book clearly depicts the things that you've endured. It certainly has opened my eyes to fully understanding the magnitude of everything that you have faced both in your childhood and adult years and the things that you have had to come to terms with. Your strength and perseverance are only some of the great qualities you have that I admire. I hope your story will help others as I know writing it has helped you. You've gone through great lengths to do whatever you needed to be mindfully healthy. This book was an emotional roller coaster and brought tears many times, a story I couldn't put down. Thank you for sharing and know that many people stood behind and love you."

—Darlene

"What an amazing story! The power of two wonderful people, two friends of mine by the names of Paul and Lorraine Hebert. He is a hero and she is an angel. I had three safety meetings this week and I quoted Lorraine each day. She said to everyone, *Don't be so selfish as to think that your incident only affects you—It affects your family, friends, co-workers and many more.* I had to repeat it. You folks are my inspiration to be a safety guy and help spread the word."

—Barry Hannah C.E.T.
ATCO Electric Ltd.
Health, Safety & Environment Coordinator

"Paul Hebert's autobiography is more than an account of his life—it reveals the inner monologue that many of us choose to ignore. His conversational writing style lends to a sense of having a best friend's advice when one is struggling to come to terms with the past and to illuminate how this determines our future. His tone is never condescending or lecturing rather he readily admits his own foibles in an effort to foster understanding. His willingness to disclose difficult truths about himself allows the reader to recognize similarities in one's own life and hopefully, provides an opportunity to choose a different course moving forward. The author's voice communicates love and patience, acceptance and an interest in catalyzing change for those that are still burdened with past trauma. The details he provides about his own life draws the reader into his story, opening the door to happiness and growth, should one then choose to put these past traumatic burdens down, and pass through to a more joyous, honest and loving way of being. The reader bears witness to his pain and heartache and shows how he forgave himself and others, which clearly communicates his eagerness to help others do the same. His discussion of his life, both personally and professionally, shines a light on the path to recognizing how these events shaped his life, providing a lens through which we see ourselves, facilitating successful navigation of the relationships we hold most dear."

—Alonzo Mitz
Love and respect always
Former NFL defensive end, seven seasons, with the
Seattle Seahawks and the Cincinnati Bengals

the Sun Above the Clouds

the Sun Above the Clouds

An Autobiography

Paul Hebert

with Rodney Miles

HEBERT
PUBLISHING

for Lorraine

The darker the night the bolder the lion.

—Theodore Roosevelt

CONTENTS

PART ONE: TOUGH AS NAILS

[1] WHERE TO PUNCH

I GREW UP in a different economy. Survival was the priority, and as a result people had certain survival skills back then, skills that I think are really lacking today. One learned how to survive on his or her own, and it was tough. In school, we focused on the basics—reading, writing, arithmetic. There was little if any government support through hard financial times. And we grew up with our parents teaching us their way of doing things. Even their emotions were passed on to us, and that's true for all of us, I think. Our parents' manners and communication skills are passed on to us, whether we like our parents (and those manners and skills) or not. That's not necessarily good or bad, but my parents had it really very hard.

My dad was born in 1905, and my mother in 1920. My father is of Acadian descent, and he came from the south of Montréal in St. Constant, along the Khanawake Indian reserve in Quebec. When he was young, he lost his little brother, Paul (who I am named after), who was only five. They were playing in the waters of a small river across the road from the house, just kids having fun, but they got wet and cold, and little Paul got sick with pleurisy[1] (which is even worse than pneumonia) and soon died. Back then they had no antibiotics, in fact they didn't have anything to fight infections with. My grandmother was so distraught, she blamed my father for Paul's death. And according to my aunt, my father was then physically beaten by his mother—beaten and abused—*every day*. It's hard for me to imagine that, and it took me a long time to understand why my own father was so mean. It's hard to imagine growing up as he must have, with such guilt, and to carry that for the rest of your life. How do you forgive yourself? My dad needed help but

---

[1] Pleurisy, also called pleuritis, is an inflammation of the pleura, which is the moist, double-layered membrane that surrounds the lungs and lines the rib cage. The condition can make breathing extremely painful. — http://www.webmd.com/lung/understanding-pleurisy-basics

there certainly was no "anger management" back then. In fact, in those days, there was no help at all.

But Dad was the oldest of five children. He used to talk a lot about how he loved school, I think because he had to quit in the fourth grade, at just nine years old. His mother died of pleurisy not too long after his little brother Paul had. Now, in those days there were no social programs—families looked after families—and my father, being the eldest of five kids, was now the one in charge. With all that pressure this must have put on him, as what we would consider today a *little* boy, I can't even imagine what he must have gone through.

I have been much more fortunate to have gotten the help I needed. In fact, the most powerful tool I ever acquired and learned to use was the ability to forgive someone else, and the person I had the hardest time forgiving was *me*. Drinking and self-hate affects everything. Your self-esteem and confidence become shattered. I used to walk into a bar or dance hall and look around for the "bad ones" because I didn't want to get myself in trouble. Later on, when I realized I might instead look for the smiling faces, things changed a lot. Stay away from the angry ones, look for the nice people. Simple. People are just like animals—stay away from growling, barking dogs, and go instead to the dogs with wagging tails and happy ways. One approach is to look for more happiness, and the other is looking for a fight.

My father, Gustave, at about 32 years old, around 1937.

Yet, we all start the same, don't we? I keep this in mind when I meet people, and in all my dealings with people today. We all have a story, and they are not all peaceful, happy, easy.

Nonetheless, with understanding and the right view, we can see that we all have certain things in common, which is a great (and perhaps the only) basis for human affairs. And I can understand, today, why my father was so protective over us. He didn't want anything bad to happen to us. My father hardly ever talked about losing Paul with us. You could see he wasn't comfortable talking about this, even later in his life. Even after all I've been through, I just cannot imagine the mental toughness they needed just to survive.

I think about it now because Mom and Dad came from the same background, and I'm not so sure that was a good mixture to have, that kind of mental toughness on both sides. And Mom was even tougher than Dad! She was a kinder person than Dad, but you didn't want to bring her toughness out if you could help it. I have not seen my mother angry many times, but she was one tough lady and when I did see her angry, once was enough! Yet she didn't use anger to empower herself. She was both powerful and patient, and I wish I'd gotten more of that patience from her. Mom was actually a very understanding woman, our mentor growing up, and I believe this is why we all survived the hard times as well as we did. You don't always realize it at time, but we were very blessed to have a mother who could be so kind, patient, and still so tough both mentally and physically when she needed to be.

Dad "self-medicated," let's say, with alcohol as we were growing up, as an escape. He would go to town and come back "over-medicated," and in this state his temper would come out, which was not pleasant at all for us kids. We were afraid of him like that. And it's still hard for me to say this, but Dad used to abuse Mom. We were so afraid for Mom because we were all too small to help her, but one day even that changed.

My oldest brother, when he was 15 years old, had already been made tougher than nails by Dad. One night my dad came home from a dance with my mom and a fight broke out between them. But this time, my brother intervened. Now, my oldest brother was so strong and tough he was known to knock people out with a single punch, and I am not kidding when I say that. I have seen him do it, actually. And on that night, thanks to my brother over-powering my dad, it became the last time we had to worry about Dad abusing Mom like that. Honestly, I was relieved Dad's beating Mom was over. And this is how we got to be so hard, ourselves. It's really true that if you treat people like crap, what goes around eventually comes around.

As he got older, and maybe to take out his aggression, my father started boxing with the Mohawk Indians in Montreal, Canada, where we were. My dad used to talk a lot about his Indian friends. They were nice to him and he found acceptance with them. I heard from my uncles and Dad's Indian friends on the Khanawake Reserve that he was a very good boxer, but I don't think he had been a street fighter. And as kids, he taught us how to box. He would stuff winter mitts as boxing gloves, get down on his knees, and teach us to box. He wouldn't punch us, he just showed us the moves and footwork we needed. He taught us to go for the middle of the opponent's chest with a solid blow to the heart, as it stuns your foe even more than hitting them in the face. In fact, he showed us that when you hit someone in the face it really hurts your hands, and anyone watching sees the damage you've done, which angers everyone. You don't want to disfigure anyone, because that anger stays for a long time. And after a fight I never felt good at all, as it was.

It's still hard talking about my father because he was a very aggressive person and it made us all tougher than nails, but that's the way it was. It wasn't an easy way to grow up. People, however, sometimes need to know that they are not the only ones that have lived through something like this, and some people I know had it even harder. I know a very nice lady who was sexually abused by her family for years and had to learn how to deal with such a past. She has a child of her own now. I asked her once how she has been able to survive. She told me she was a mess as a child, and that she now has PTSD, but she has also acquired survival skills, and she's a very nice lady today because she was never a bad person to begin with. Bad things happen to good people, but we can still be good people.

Mom (Jacqueline) seemed to have had a similar life to my dad. Her parents moved the family to Northern Alberta in 1927 from Quebec City, and her father was a policeman with the City of Quebec. Everyone was after free land at the time so my mom and her family moved to Kathleen, Alberta, where they claimed a homestead—a farm, but my grandfather had no idea how to farm or what it took, and the whole family suffered for it. I believe there were times so tough they didn't even have food to eat, and I believe it happened more than my siblings and I even know.

Her own mother, my grandmother, passed away in childbirth, I believe when my mom was only 10 years old. So, like my dad, my mom also had to stop going to school because she had to become the mother, the caretaker of her five brothers and sisters. Can you imagine having to do this? My God, they were just children themselves, raising children. And all at a time when you did laundry by hand. I can't even imagine the hard work they did, all while having nothing we today are accustomed to—no electricity, no fridge, no appliances at all. In the winter, they burned wood for heat. And to get the wood, you had to chop down trees, of course. You milked cows for your milk. You churned your own butter. If you were fortunate enough to afford coal oil, you had lamps. I mean, my mom and dad really had nothing, but they really did have survival skills. And they knew hard work.

My mom, Jacqueline, I would say in her early 40s, a very pretty lady, a great lady, in fact.

My mother left home at the age of 13 and I actually don't know why. From what I have heard, though, it wasn't good. I remember one of the neighbors saying that they saw my mom walking away in the snow with only leather shoes on her feet, in the ice cold of winter. They stopped and gave her a ride, and they brought her to their place where she later worked for them, for the Dupuis family, who had a sawmill at the time. So, at 13 years old my mother was employed, cooking and cleaning! This was in about 1933, coming out of the first depression, when the markets all crashed. This was when most people in the cites lost their jobs. Then the Great Depression pushed people from the cities back to the country because farmers, of course, grew their own food.

Survival skills became very valuable. People were grateful just to have a place to eat and sleep. No one today seems to understand economic disasters like the Great Depression, but the way I see it, the stock markets and the banks had *spent* everyone's cash, and today big corporations and banks are regulated by the government to keep a certain amount of cash on hand. The first government assistance appeared back then, too, called "relief," which was barely enough to survive on, and a big embarrassment to accept. There was a lot of poverty back then and it took a long time to turn things around—until the Second World War, in fact.

Seems like today, we might not realize how *recent* such hard times were and what they actually lived through, or how necessary the skills were that they took for granted, that so many of us today not only lack, but might find hard to imagine. Mom wouldn't waste anything at all—to waste food was a sin, in fact. My dad was the same way—you didn't waste food at all, and if you didn't clear your dinner plate you would hear about it! They would even save paper and cardboard because you would light fires with them, and without them you had to make wood shavings to light fire with. Back then we didn't have plastic, either—we used glass or granite cups.

In today's society, we have government social services to help us. Back then, we learned as kids how to grow our own food! I still know and remember these skills today. And what if the government runs out of money someday—then what? It's no wild idea that there may be hard times ahead, as our governments spend themselves broke. In the recent past, only the strong survived, but more of us had survival skills. There were no social programs, and today we see our social programs, again, *going broke*. If *these same* survival skills are not what's greatly lacking in today's world, we certainly need to educate and equip ourselves with much better skills for living, otherwise.

It helps to have friends, to have them and to be one to other people as well. I can talk to my friend Jim about anything if I need to. In fact, Jim bailed me out (financially) at a critical point in the past (and I paid him back), and I am still grateful—to this day he's still my best friend. But there are people who will borrow money and not repay you, which is just not right, of course. I have had people do this to me, as if they don't realize it's a breaking of their word, and if your word is no good, what do you have? My daughter Courtney, like most kids, has borrowed from time to time and I don't mind because not only is she my daughter, but she has always paid us back. Courtney has a *conscience* and she is known to be honest, which is a very big deal because when you are honest with yourself and others, you can live with yourself. Pretty important.

A proven approach to lending, according to my friend Monica, is when someone asks to borrow money, take your wallet and look inside, and ask yourself "How much can I *give away* today?" See, lots of family loans never get repaid, and it would be better to *give* what you can away than to see familial relationships suffer over money. If they can't go to a bank or credit cards and borrow, they are likely tapped out—leveraged to their limit! And there is a point where "helping" them with a loan is *enabling* them—or rather, *disabling* them. Don't let friends and family go without food, of course, but handle things in such a way that the important people in your life find their own feet. Be an enabler, not a disabler. It's called "tough love." But know that telling someone in need you will give advice for free doesn't go over very well! A friend came to me once who was having a financial problem. I told him, "Write down everything you spend each day on a note pad, in fact, set up three boxes: The first box for unpaid bills, the second box for paid bills, and box three for how much money is left over. This is basic accounting." It was not exactly what he was looking for, and he went to his banker who told him basically the same thing, to write down what he was spending each day.

The government system should be fairer, too. I don't think it's all bad, but there has to be regulation put in place to protect investors, for example. And as good as regulations can be, education is more important. Even without good regulation in place, you should educate yourself somehow on financial matters because to get involved in banking and stock markets you have play smart. And "playing" the markets is exactly what it is if you don't know what you're doing! Playing at the casinos is the same thing, really. After two serious losses in the stock market of my own, I understand it a little better. I lost good money after I invested in precious metals, and it took two years to gain it back—I was lucky and fortunate at that, actually, and I learned a lot from that experience. Utilities might be a better investment than precious metals, for example, because people never use less electricity and the price never goes down. There are good investments out there, but you have to find them. And no matter what your plans are, if you started by cutting up all of your credit cards, it would be a great start, and a great feeling, actually.

It's too easy to rely on the systems we have in place today, and I think we lose a lot because of it. Life might be richer were we not so dependent on the financial and government systems we have in place. It's too easy, too encouraged, in fact, for us to get deep into debt, and to then get into trouble of all kinds if we should fail to meet the demands debt places on us. Survival skills like we used to need would be useful to know and practice, perhaps, as a way to keep us tuned in and appreciative of what's important, of the people around us rather than the things.

[2] THE LITTLE TREE

MY FATHER USED TO WORK with some of his Iroquois Indian friends on trashing crews, back in 1925 or 1926. And as a kid, I used to ask a lot of questions about how my father traveled because I wanted to travel like he did, someday. I used to ask what it was like, riding the train in those days. I remember he said he had gone to work in northern Ontario in logging camps. They used to cut down the trees by hand, with a method they called the "push and pull." Two men would push and pull the long saw (and if you both pushed at the wrong time the saw would bend), then manage the falling trees by hand, roll the logs onto a platform, and then haul them away by horse—in fact managing the animals was just as much a part of the skillset they needed back then. Later on, in the 1960s, we got our first chainsaw. Well, we didn't know anything about chainsaws— sharpening or using them—so we learned by reading the instructions and by trial and error.

My dad told me he had traveled to Western Saskatchewan, where he went and worked on trashing crews south of Regina, a small town of Wilcox. He said it was like an ocean of wheat, waving in the wind. He loved it. He wanted to homestead[2] in Saskatchewan but there were no more homesteads available. He tried instead to get homestead land in Castor Alberta, but again, none were available. To homestead was a big deal because the first on the land would become a pioneer and own the land they claimed. It was a very exciting thing for people, and to own your own land, a dream come true.

My dad's relatives back home told him to go to Northern Alberta, because there was land available, so he traveled to a small town called Falher, Alberta, a small French-Canadian settlement in 1928, and he finally got a homestead of over 160

------

[2] "NORTH AMERICAN historical (as provided by the federal Homestead Act of 1862) an area of public land in the West (usually 160 acres) granted to any US citizen willing to settle on and farm the land for at least five years." —Google

acres. He said all he had at the time was 50 dollars, his World War I pack-sack, and a few important things like an axe and a shovel—everything he needed—that and the clothes on his back. There was an old shack on the land that a former homesteader had started and left, a rough-lumber shack. The only insulation was tar paper on the walls and it had a wood stove. They truly lived off the land. My father cleared 15 acres that first year, and another 15 acres the next. Imagine clearing 15 acres by handsaw! He told me they would burn the forest first, to reduce what had to be cleared for farming. That's a lot of work, and the land had to be ready to seed the next year.

In 1930 he borrowed $400, which was a lot of money back then, to buy a team of horses. Again, this was 1930, when the depression started. Dad was growing oat crops, but oats only sold for seven cents per bushel, which barely paid for the freight on the railroad. My dad didn't have it easy but this was the way everyone had it in those days, all in the same boat. It wasn't easy for anyone, but they were young and they understood they had to work to get anywhere—nothing was handed to anybody for free. For extra money, my dad worked for other farmers too, clearing land. This way he could make his payments to the bank. I know this because I used to ask my dad questions for hours. I thought his life was so interesting. I think this is how I connected with my dad and fell in love with history at the same time.

My dad, brush cutting trees in the early 1940s with a John Deer tractor with steel wheels and a homemade brush cutter.

They were never bored, even out in the middle of 160 acres. They used to get up at four o'clock in the morning to feed the horses and begin each day of labor. The whole family did as such, that's the way it was back then, and they did very well with what little they had. It required diligent management of affairs. When you have little or no means, you have choices but absolutely have to live within your means. This isn't completely the case these days. My father used to say, "It's not the bad times that break us, it's the good times when we buy too much." A prevention they figured out was to put some away for "rainy days," to "make hay while the sun shines," and to appreciate the good times. It took me a while in my own life to really understand what they meant, but now I do.

When we are young we live more in the moment, and we like instant gratification. As we get older we start to look more at having a fixed income, because we don't want to work forever. You want at some point to enjoy what you worked for all your life, to stop and smell the roses, so to speak.

Beyond the hard work there was some happiness. They were not bored because they knew how to entertain themselves. They played card games, played ball, held dances, and played music. We lived half a mile from Lake Maguire and there was a one-room country school and a little store with a post office across the road from the school. On weekends, they used the school as a dance hall, where they all had a lot of fun. Dad used to walk six miles to town, to Falher, every Sunday for church and to socialize. They would visit there and in the summer months they played ball after church. There was, of course, no television, in fact they had no radio back then. What they did do was play instruments, sing, and dance together. But they were socialized, they learned to interact with their neighbors, and they were never bored.

This was true for us as kids also, because if we had no work to do, we would play sports. We'd play baseball without mitts—we'd just grab a bat and a ball and go for it. We did the same with other sports—hockey, for one, with homemade hockey sticks and an old hockey puck that had chunks missing out of it, but that old puck was everything! If we lost it, we would look for it until we found it because that was the only one we had. I rode horses a lot, and we laughed a lot together. If we wanted to skate we had to clear the skating rink with shovels, which was a lot of work. It would snow a lot, and by the next day the rink would have to be cleared again, so we played a lot of football in the winter months.

With my brothers and sisters. Look at the big smiles on their faces! I'm the smallest and the youngest and have that same big smile today. I survived thanks to humor. If you can laugh, you can live through anything.

Today I still stay busy. I don't like to sit around at all, and I think it's because we were so active as kids. I golf a lot. I used to hit a lot of balls at the driving range, but I have slowed down some lately. I used to walk nine holes of golf, too, but I had to slow down because of a sciatic nerve. Nonetheless, I go to the gym as much as possible. I try to *live* reality and not just watch "reality" on TV. I love my life and I want to live it the fullest I can, and a busy mind is a healthy mind.

Dad built a very nice farm in the 1940s through the 1950s. We had a cistern—a concrete tank for storing water—and we had a 32-volt generator and batteries, enough power to last us about a week. We had a radio, and due to limits on power, had to limit listening to it to just a few hours per day. But growing up in Canada, we *never* missed hockey night—I remember the "big six" NHL teams and most of the great players. When it was very cold, we would listen to hockey on the radio from Edmonton, or on our local station in Peace River. I remember this like it was yesterday, and it feels good to remember those great days.

We got electricity in 1957. Back then, in cooperation with rural electric associations (or "REAs," ours was called "Jean Cote REA"), farmers cut their own power poles from pine trees and dipped them in tar-creosote. In fact, forty years later I would change a lot of those poles myself, after I became a lineman. But when we first got electric, I was seven years old on the farm and the Hachey Power Line company came to hook us up. I remember the linemen climbing the pole, one of them a native Indian named Sam Gladsoe. I saw him again, many years later, after I became a lineman myself, but as a kid, it just amazed me, watching them climbing that pole. I think that was when I realized what I wanted to do, it just took me some time to get there.

In 1961 we got a television, and we had *one* channel. We didn't have a phone until 1967, but my God, how we did communicate with that thing! If you lived at any distance from someone you would write letters. You'd ride your bike to see friends (mine were about four miles away) or walk if they were close enough. On rainy days, we would get to sleep in and rest, It was nice doing nothing once in a while. I remember one day, when it was cloudy and overcast, an old Indian friend named Terence told me, "Remember, it's always sunny above the clouds." He said the trees in the forest grew straight and tall, that they shot up for the sky reaching for the sun. But there was a little tree, growing very slowly, and not straight at all. One day a strong wind and heavy rain came, and all the tall trees blow over, except for the little tree, which was left standing, undamaged. See, the little tree was growing roots, and all the tall trees were growing straight up, shooting for sun, too busy for roots, too rushed to be flexible.

Fewer options make your choices easier if not obvious, or you might have no choice at all. For my parents, it was "survive or die," and "die" was of course no option at all. The native Indians we knew, they had survival skills and no money—their community did not use it, actually. They learned from their elders and lived off of the land. They picked wild potatoes, wild carrots and blueberries. They would dry moose meat and mix it with fruit. The vitamins from the fruit would absorb into the meat, and they buried the meat in the ground, to be dug up in the winter. It was very healthy for them, and they didn't get sick like we do now. They, like my parents, got tough—tough as they needed to be, and I guess few of us get tougher than that.

[3] HAPPY JUICE

M Y ONLY WISH is that my siblings and I could have all stayed closer to each other, distance-wise. We were close as a family and now, naturally, I suppose, we've grown up and drifted apart. One of my brothers passed from cancer at just 50 years old. I miss him, and I miss time with all of my siblings. We were all raised under very strict discipline, and it was stressful. But my oldest brother, my God did he ever work hard and get the harshest discipline, more so than any of us. He was very protective over us, his brothers and sisters. I think the physical discipline we all received was way too hard, but we lived through it. In our house, you didn't dare make the slightest mistake because you would hear about it quickly. Years later I would finally resolve to *forgive* it all, because the anger I walked around with was killing me. Forgiveness was ultimately the only option to rid myself in a large way of that anger, I just couldn't live like that anymore. But when I did forgive it all, I can tell you I did find peace with myself. The power of forgiveness is so strong. "Therapy" back then was limited to locking somebody up in the "nut house" only after they had actually injured someone for mental reasons.

I know that my father abused my mother, but to what extent I cannot say. It's still hard to write about, to talk about, it still hurts to think about. Mom was very quiet about it all. Today, I do talk about this with my own psychiatrist, but beyond that it is still just too difficult. And the abuse *we* suffered is still just too hard to really talk about. Humor and survival skills, I believe, are two things that have helped me (or anyone, for that matter) through, and through to an active life filled with happiness. Somewhere the chain has to end: My father was himself abused, he abused my mother, they abused my older brother and us as well. I used to blame my dad, but he went through the same, and there was an understood (if not completely believed) argument that this was "good" for you, that it made you tough, that it added to our chances for survival. We know different today, so the trick is to separate out a robust and direct approach to life from the ills that are disguised as "beneficial."

Before I developed any coping skills, my response was simply anger—anger I carried with me everywhere and into every situation, like a cancer or a poison. It's important to see how this anger prevented me from really sitting back and assessing situations objectively. I would get angry. I would remain quiet. I drank. I would let the situation swell until almost out of control, at which point I would either explode or implode. People feared and avoided me—you can approach a dog who wags his tail, but you avoid the ones that growl, of course. I only, previously, got far enough that I might warn someone if they were bringing me to such a point.

It's in our culture. I recall people who would literally get into fist fights every weekend, as they "needed the release." The bigger ones pick on the smaller, and so on. I didn't realize initially that this was a form of *defense* for such people, but I see it now. And I see the value in staying calm, staying friendly, or at least, staying away. It's simple, really—look for the nice people. Look for people who smile, not scowl. Seek out happiness in others and in the world, and you'll be happier yourself. And if you have to work with unhappy people, at least keep things as simple as possible. Don't make things personal and stay out of personalities as best you can. That's where people get quickly threatened and react rather than act.

I remember one fellow, after having worked with him for about a week, said to me, "I do not like you at all."

"That's okay," I said, "you don't have to like me. We only have to work together." He looked at me and he didn't know what to say. "I come here every day," I said, "just to do my job, that's all. I don't come here to make buddies. I need this job, I need the money. I will stay out of your way, and in return you stay out of my way, and we will just do our work."

He looked at me for a moment and then said, "I think I do like you."

Principles before personality has saved me, more than once.

I used to have a "sheep" mentality. My dad used to tell me, "Never think you're smarter than your boss!" So, I didn't, even when common sense told me otherwise. "Do as you're told," is another one. Often as children we look to our parents and our teachers to *tell us* who and what we are—a "good kid," a "good student," or even a bad one. Both are intimidating, and that's probably the point. In fact, things like that are not just used to gain obedient children but to create control over a worksite. But it robs a person who believes it of their common sense, of their autonomy, of the *self*-evaluation—which is so important—and to a good degree, of their ability—their critical ability to observe and make decisions. Unsafe, actually, and short-term thinking.

It suggests to someone that they don't have any common sense and need to subscribe to rules over reason, to not *think*. But common sense is derived from common practice. If you have performed a job task over and over again it has become common practice. Young people often have not yet achieved common practice, so we have to teach and guide them at first. It took me a long time to learn this, but there are good ways to work with younger employees, and most of them want to do well. But to say or suggest that someone has no common sense in this way is not guiding anyone at all. I know, I have used it in the past because people have used it on me. All it does, actually, is set someone up to fail, where people, especially on dangerous jobs, need to be able to observe and make their own decisions based on both what they've been taught and what they see.

But if you are big and strong, you can intimidate others and become "the boss," at least until some other bully shows up. Some enjoy the power trip, which is worthless. I like working with people to bring their positives out. Fear and anger are a nasty combination at work, in dangerous jobs, and of course, at home. You can create too much adrenaline. But if you empower others, you produce what I call "happy juice," which comes from playing sports, for example, and from feeling good about yourself. Give compliments when they are deserved. Make people feel good about themselves. What you start usually continues. When managing a job site, all I really want to do is make things easy on myself. If I make everyone angry, that is very hard to handle. You do not want to lower everyone down to your own bad attitude, yet this happens more than a lot. People react with emotions, and a good, happy leader considers this before managing others.

Many managers use fear as their main method of gaining compliance and control, but fear backfires. Fear creates an unsafe work environment. Fear leads to group discontent because people talk. Men, on a work site, talk as much or more among themselves than a sewing circle does. But all you need to do is listen. When someone talks to you about a situation, realize they often do not really hear you because they are too tied up in thinking about what *they* are about to say. You need to re-focus them somehow. And that applies to you, too. What people tell you is usually more important even than safety audits and paperwork. A real leader who creates a safe and easy work environment listens patiently. When you do this, when you are nice to people as often as possible, and you listen attentively and provide firm, positive leadership and guidance, people under you start coming to you for answers and leadership. People recognize good leadership and depend upon it, and they hate being pushed to work through intimidation. It doesn't work for all people, but for most, if you are good to them, they will soon be good to you. And if someone is very stand-offish, just give them the room they need.

Tell them what you want done then help them if you need to. If they are not capable of doing a job on their own resist the urge to micro-manage them, as this is the worst thing you can do. With micro-management, you are slowing the progress down in the long-term. Let them learn on their own as much as possible, because those lessons are the best-accepted and the longest lasting. You are the supervisor. You guide them. This is your job. You are the coach, and when they need your guidance simply give it to them. Compliments well-deserved make them feel good and empower them—possibly the best tool you can use.

People need to learn to use their own brains and to trust themselves before they can make good decisions, so we need to empower that. People who are abused, physically or psychologically, do not respond well. They are defensive. They exist in great number. It takes time to work with them at first. Their self-confidence can be non-existent. Emotions are a big problem at first, and they have learned to use their emotions for everything, as this is all they know. They have been controlled too often in the past, and with bad results, and they know it, so they resent it. Working with them takes a lot of energy. We have to work with them one-on-one. Be positive and explain to them exactly what you want. Personality has nothing to do with the job you want completed, so just keep conversations simple and stick to your plan. Finding ways to work with all different kinds of people is very important. We are all unique and need to be treated that way.

As a young person, I started off not doing so well at school. I couldn't keep up with the rest of the class. I was near-sighted because of astigmatism and needed glasses in first grade but I didn't get them until the end of second grade. I had a hard time with exams because I would go into a panic, I was so afraid of failing. And my teachers didn't know any better. Back then you were simply either smart or stupid, there was no in-between at all, no knowledge of learning disabilities. Today we seem to have all kinds of solutions, but I still go into panic mode if I have to take an exam of any kind. Even if I recall what I studied, I still go blank on the test.

But as a kid, I didn't understand what was going on. Other kids would laugh at me. I was called names like "stupid," and "dummy," and I didn't know why. In fact, other kids would actually poke at me, often until my confusion turned into anger. It was a defense I resorted to because I didn't want to be treated like that. My anger, though, led to my getting back at these other kids, which led to my father punishing me with the strap—so often that it started to not bother me, I started to be able to hide the pain. It did teach me that we might be able to mask physical pain—to not even be bothered by it, really—but we cannot so easily hide or ignore or forget the emotional pain.

As children, we often do not know how (or why) to forgive, and even as adults this is often hard or impossible. It's important, though, to get a handle on your own desire for revenge, a handle on the subject of forgiveness of others and forgiving yourself, because without those things you become the victim of your own anger, fear, and resentment. It's you who has to be with you all of the time!

So as a child in school, I found myself seated at the back of the class by the teacher, due to the problems I seemed to always be a part of. That might have solved the teacher's dilemma, at least in part, but I was still confused about what was going on with and to me. People minimize others to "maximize" themselves, and it results in the many stereotypes kids and adults alike throw about. The irony is that in the real world, you cannot actually weaken the weak—they are already weak! All that does happen is when someone tries to use his or her own anger to empower themselves, which is an illusion, you actually create more long-term problems because there are those, like I used to be, who will not forget or forgive. There are better ways to handle all sorts of things, positive ways. The good feeling that seems to come from revenge, from using anger to get back at people, is really not "good" at all. It means you are not using your resources, that you are actually a kind of lame duck. When you force people to instead help each other, some kind of education actually takes place that does not take place with a battle of force.

Here's a trick I've learned in managing people who are not getting along: This works in work, at school, and in personal life. Take the two of them and assign them to help you. Ideally one will be more experienced than the other at what you need help with. Tell them you need help and then bring them both up to speed on the task or project. Set them up for success by providing clear objectives and whatever they might need to accomplish them, and encourage them to work as a team, offering both reward and recognition for success and positive, constructive reinforcement. As they start accomplishing even small things together they will start to bond and improve at the same time. I actually know this from experience. When I was in the fourth grade, there were four of us classified as "stupid." As a result, we started to get along with each other. We started hanging out and having fun together, and we started improving in school, too.

I used to box a lot with my older brother and as a result, had gotten pretty good at it. I remember one kid who started picking on me and just would not stop, so I beat him pretty good—I had him on the ground pretty quickly. "Kids will be kids," right? Well, you find out after a while that this all just makes everything worse. People who attack build defenses. People who get attacked build defenses, too. Both build emotional defenses. What did Gandhi say? "An eye for an eye leaves the whole world blind." When people spot weaknesses in other people and exploit them, it creates a lot of bad adrenaline, and when "high" on *bad* adrenaline, things get dangerous quickly, whether it's work, school, or play. Nowadays we have laws against assault and even bullying, but it still comes down to a matter of understanding that neither to bully or be bullied are desirable. Often this is very subtle, too, as when someone plays off bullying as joking or kidding around. Still toxic.

So, finally understanding this, I took a good look and realized I had a tendency to be *the bully*. I had developed that aggression as a form of self-defense, and I believe it came out of my younger school days, as a kid who got picked on verbally and physically for being "stupid," when it really had to do with poor eyesight. Later in life, I simply decided I wanted to be a nicer person. It took discipline! But you realize (hopefully) that no one has to live with you all the time except you! You can make friends easier than you keep them, sometimes, unless you are a nicer person at heart. To combat being picked on, I became the tough guy and picked on others. It did result in better treatment, in getting *respect,* but I later understood respect out of fear is a different kind of respect, and I decided to change. Fear-driven "respect" is what I had, and I would learn eventually that you cannot control another person, ultimately, with fear. The results are temporary at best, but the negative effects are long-term. And by poking at someone's weaknesses you might feel tough, but you eventually, if not right away, bring out the worst in people. What choice do they have?

To make a stronger team, help the weaker ones *with* the stronger ones through group and team activity, and through positive reinforcement.

[4] SMALL MIRACLES

BUT MY HISTORY WITH VIOLENCE was not over quickly as a young man, not by any means. One time, two brothers got a hate on and tried to hang a good one on me. It didn't work out very well for them, however. I licked the younger one, then the older one, then they both tried to take me at the same time. Still didn't work out very well for either of them. One of them grabbed me from behind and punched me, and I felt no pain. The next thing I remember, I was a wild man, and they were bleeding. Gus, a friend of mine, saw the whole thing. "Man, they never even had a chance, you were all over them!" he said. And I was.

Good thing he was there, too, because Gus backed my side of the story and saved me from a bunch of trouble at school over that one. And even though I was the "victor" in that exchange, it was still traumatic for me. It was at a time when I would get hit and rather than feel pain, simply go blank, go wild, go black, start shaking, and implode. I had "survived," but not very well, by my built-up defenses.

It was my dad's beatings that had gotten me to where I would no longer feel any pain. He made me tougher and harder, something lots of parents at different times and places have done to their children, some by design and some not so much. Once my dad started beating on me he wouldn't stop, and I learned to block out the pain. My dad would beat me until I passed out, and more than once. As hard as this is to talk about even today, and all these years later, I have to believe telling my story will help someone else. In my case, the only way I eventually got past all of this was to *forgive*. That's usually not someone's first choice, but that's what ultimately set me free of the emotional damage Dad had done. The person carrying around the resentment and all of that hate, after all, was me, regardless of what Dad had to live with in his day.

And before I could forgive Dad I had to forgive myself. See, I had become just as mean. The anger and the resentment were running my life. I was negative, I was unhappy. I hated myself for how I treated others. I found myself with no friends, but what I did get out of it all was what I learned. First, I learned that my dad had lived through the loss of his mom. He had also lost his little brother, which he was blamed for. Before knowing this my sister and I grew up hating Dad, hating him until our aunt explained the hardships he had gone through. It at least helped me understand what he was going through. It was the start of my making peace with my dad, who suffered a stroke later in his life. I remember going to see him in the hospital. As I walked into the room he recognized my sister and me, and he started crying when he saw us. He couldn't speak because of the stroke—he tried, but he just couldn't get to the words. We figured out he was trying to say he was sorry, I could see that. In response, I said I was sorry, and that I wished I would have done better growing up. He cried a lot then, just as I am crying now as I write this.

Forgiveness is *so* powerful.

I realize now how important it is to both make amends and to forgive. It's healing, it's uplifting, and I felt like I had healed my soul that day. We all had been through such hard times together, sometimes we took it out on each other. And I had not realized my dad loved me—that's not what I used to think when I was a child. So this was a big healing for me, going directly to a person and making amends or forgiving them is very powerful. But there's a way to do it. If I have harmed someone, I have to forgive myself first and be willing to accept *not* being forgiven by the person I harmed. To be able to do this, I had to change my mind, I had to change the way I was living, and this helped me understand myself and others, too. But I can live with myself now. Forgiveness put me at peace with myself.

Forgiving is learning to love yourself again, and that was the hardest thing I think I ever did for myself.

Between the physical and mental abuse from Dad, the hardships at school, and other things, I was a very lost person. I didn't know which way to turn. The only person that could comfort me and my siblings was Mom. She would talk to us. She was such a nice lady and she had such a hard time. She was such a good listener. She would make us feel good by hearing us and she would reinforce that we were good people. I know if we would have lost her, we would have had a lot harder time growing up. My dad, as mentioned, was not nice to my mom, thanks in part to his own anger issues. It's sad because he never, ultimately, dealt with his problems. My mom was abused as a child, went to work at the ripe old age of 13, and married my dad when she was only 16 (Dad was 31, which I think is too much of a difference in age). They worked hard on the farm, often from daylight till dark.

But hard times are a great teacher, and I learned a lot from them. I learned to deal with my problems. I remember in seventh grade there was a teacher named Mrs. Antil, who was a great teacher, and I remember walking into her class and going straight to the back of the room and sitting down. See that's where I had been sitting in my other classes, and that was where I was most comfortable. She called me to the front of the class, and she told me to sit in the front there, right next to her, and I felt so out of place. At the first break, she told me to stay seated, and she came over and started talking to me.

"I want you to sit in the front of the class," she said.

"I'm not very smart," I said, and she looked at me in a way I had never had anyone look at me. I was very uncomfortable.

"Paul," she said, "you are not stupid at all. In fact, how do you think you have made it this far? You have learned everything by memory! You are *not* a stupid person. I am going to work with you."

Mrs. Antil was a great teacher. She helped me learn at my own pace, and I learned more that year than I ever did before. I started scoring better on my exams. My self-esteem was better, but I still had problems. She told me other kids had better memories than I seemed to, and she taught me skills on what to do with exams and how to go through and answer questions so you don't get stuck on the ones you don't know (just keep going and answer the ones you do know, then come back to the others later, if there's time). She made things so much easier for me. She has passed on now but she helped me more than I even realized at first. She was a small miracle for me, maybe even a big miracle. I think of her often. And the little confidence she gave me has stayed with me for a long time now. We all have to learn at our own pace, and ideally with our own purpose. I did go back and see her years later, in 1984. It was nice seeing her. She was the best teacher I ever had.

But when school was finally over, I was happy and relieved. I ended up leaving school at the end of seventh grade. I was 14 years old and suddenly struck with appendicitis. After the operation, I did return to school, where one of the older kids who had it in for me found me and kicked me in the stomach. It hurt, but it made me even more angry. I waited for a week or so to let myself heal. I then went back and "saw" him, and it wasn't good. I ended up breaking some of his teeth, almost knocking him out. The school got involved I was asked to leave. I was happy actually, at the idea. I was off to technical school, or "reform" school as some called it. But I called it "charm" school!

When I first arrived, I had a meeting with the head priest (or "brother"). "You have to do exactly as you're told, and if you don't do as you're told you are going to be in deep trouble," he said, "and you will be disciplined according to your actions!"

What the hell did I get myself into?

Once, while standing in a line behind another kid named Kessler—I will always remember him—I saw he had a brown, lead pencil in his hand. He was looking at me and laughing, calling me names. I said nothing, I just tried to ignore him, but all of a sudden, he *stabbed* the pencil right into my leg, about a half an inch or so! I reacted. I grabbed a hold of this guy and I started beating the crap out of him. The brothers quickly arrived and stopped us, which wasn't easy to do because I had decided to take the one brother down also! I didn't even feel the pain in my leg. My punishment was to go pick rocks, and I hated how this "school" was more like a jail.

Even sleeping there, in a big dorm, was like a jail. Eating in a big, supervised dining room was like a jail. I didn't like it at all. I wouldn't play outside with the other kids. I was so homesick—I missed my mom so much, I found it hard to sleep. I felt abandoned. And for my punishment they made me pick 200 rocks and put them in a pile. When everyone else went back inside, I was left out to pick my rocks, and when they were not looking, I ran away. I had no idea where to go at first, but I had an uncle in Edmonton, so I started walking. They had police looking for me. I found the address in the phone book and I kept walking, even though I had no idea how to get there. I asked directions of an older man I encountered. I was on the north side of Edmonton and had to go to the south side of Edmonton, about 10 miles. He asked me how I was planning to get there, and I had no idea.

"Walk, I guess," I said.

"Are you from here?" he asked.

"No."

He looked at me. "Do you have any money?" he asked.

"No," I said.

He gave me a bus ticket and he told me which bus stop to go to. As a country boy, I was totally lost! But the bus driver helped me get to my uncle's. On that whole trip, I was so lost and afraid, but I was *not* going back to that "school"—I was *done* with that part of being stupid. I found my uncle's house, and I hadn't seen him or his family in a long time. He asked me what I was doing, and I told them what happened without mentioning all the details, but I felt my uncle knew. He wanted to bring me back.

"Forget it," I said, "I am leaving! I am not going back!" I just wanted to be done with that part of abuse. I would have rather lived on the street.

"Stay, then," he said, "We will work things out."

So, I did. But at my uncle's I just never felt right, I felt out of place. I had nowhere to go, but I felt I was not wanted there, either. I could see and feel it from them. But at this point I also didn't care anymore. I just wanted to be left alone, to have some kind of peace of mind, but I couldn't find it at all. Again, I would have happily lived on the streets. I wasn't lazy, so I would have found work, but I was done trying to please people, I thought. I stayed at my uncle's for just four or five days when unannounced to me, my dad showed up. I would have run away had I known, and it wasn't good when my dad arrived. We drove home the 280 miles and I got shit all the way home. When we got home my brother accused me of stealing money (I didn't) and I got a hell of a beating, again. "Since you're not going back to school," Dad yelled, "if you stay here you're going to work, but there's no pay."

When I could, I left home, at the age of 15.

PART TWO: ON MY OWN

[5] GOING TO CALIFORNIA

I FOUND WORK AT A SAWMILL. And I was so *happy,* happy to be out of there and on my own. In fact, I never wanted to go back. I just didn't want to fail anymore. I was on my own with a job. I started working hard and decided I wouldn't give up. We started work at seven o'clock in the morning and worked till six o'clock at night, with just one hour off for lunch, and we worked *hard.* I used to go to sleep sometimes as early as just an hour after getting home. I was burning out. I was young, and I wasn't very big, but I made it! I was happy. I was at peace. I was away from that hell. No one gave me a hard time, I just worked hard and that's all I wanted then. I think we were making just over a dollar an hour back then. I never called home and I never went back. My mother wrote me a letter asking me to come home for Christmas, so I did, but my dad was so mean I left. I missed my mom a lot, and I never wanted to go back *ever again,* but I was a sucker for punishment.

That January my brothers got a logging contract, so they came and saw me. They convinced me to come back to work with them, and they treated me well on the job. I learned how to operate a chainsaw (I was 16 years old) and I was as proud as a peacock. I loved it, bucking[3] logs for plywood, and I remember to this day the details we followed. I was doing something productive and I was learning something. I was on top of the world. With my brothers, I ended up back on the farm, working it. A parcel of land for homestead came available and I claimed it. My brothers had acquired an International TD-9 and they did land clearing, so I learned how to operate heavy equipment and I was good at it. I was excited about getting land and excited about working for my brothers, who arranged to keep the crops they harvested to pay for the land clearing. And there was a lot of land to be harvested and

---

[3] "Bucking is the process of cutting a felled and delimbed tree into logs. This can be a complicated process because logs destined for plywood, lumber, and pulp each have their own price and specifications for length, diameter, and defects." —Wikipedia

cleared in Northern Alberta, an amazing amount, actually. They used to use a brush cutter[4] to cut the trees down and then use a Caterpillar[5] and push the trees into big rows, then burn them all. They would then take a breaking plow[6] and break the land. All Northern Alberta was cleared this way in the 1960s.

My brothers' International TD-9 had about 35 horsepower—small, but it was a start. They traded it for an International TD-18 of about 70 or 80 horsepower, which was a lot bigger, and then traded that for a D7 Caterpillar. They then got a 1200 Case tractor with 120 horsepower—and this was big for the 1960s. It had a breaking plow, and this was a nice machine in those days. This was big equipment back then, especially when compared to clearing land by hand. I can't remember how many acres you could clear—it all depended on how big the timber was—but I know we used to break one acre each hour with the tractor, which had steel mesh tires (and

---

[4] "A power tool worn on a shoulder harness consisting of a rotary head with a small circular saw at the end of a boom, for clearing various kinds of rank or low woody growth (brushwood)."
— http://www.treeterms.co.uk/definitions/brush-cutter
[5] "Caterpillar Inc., is an American corporation which designs, develops, engineers, manufactures, markets and sells machinery, engines, financial products and insurance to customers via a worldwide dealer network." —Wikipedia
[6] "plow. also plough (plou) n. 1. A farm implement consisting of a strong blade at the end of a beam, usually hitched to a draft team or motor vehicle and used for breaking up soil and cutting furrows in preparation for sowing." — http://www.thefreedictionary.com/plow

therefore never any flat tires, modern technology back then). They used to use steel wheels on the older tractors back in the 1940s and 1950s. We worked 12-hour shifts, 24 hours a day. In 1967, we broke over 3,000 acres in 3,000 hours. I don't know if that was a record, but that was a *lot* of land.

And as far as working together and "management," there were a lot of negative controls, never giving a compliment and always barking that you could have done more. This was the way it was back then. A lot of companies squeezed production out of their workers this way, always threatening job security, pitting people against each other with statements that "the other shift is better then you are," and so on. Just negative controls. Fear was a big motivator. Anger, a loud voice, screaming and intimidating people—this was a supervisor's way of controlling the lower class.

That winter my brothers got a big contract to clear 1,500 acres of land for a big pasture. I worked night shift again. We got it finished by February and then worked for an oil company, cutting seismic line, west of Steam River, Alberta. I didn't get paid much at all and I don't know how many hours I worked, but my brothers were going broke, so I didn't push it. In fact, I worked on the farm with no pay. I just didn't even think about it because I had land and I was okay with that. But I did get discouraged at how tough my life was shaping up to be, and I started drinking a lot more as an escape. I guess I had to get rid of the pain somehow.

I left my brothers' business and went to work in Edmonton for Caterpillar in the welding shop, and I learned how to rebuild Cat undercarriages. I liked the job. I was having a good time. I partied and had a lot of fun. I was 19 years young, and I didn't get into fights anymore. I was just having fun as a teenager. But some of my family called me a drunk—the negative controls made them feel good, I guess. And I guess that's all they ever knew. They tell you that *you're* always the problem. They try to "fix" the person all the time, and that's what screwed people up—they try to "fix" others when they're the ones that need fixing. I had my own problems and when I look back, I can see the lack of social, management, and communication skills, to the point that only fear, anger, and negative controls were used—only mocking and breaking down people, which I now know does *not* work.

But I have learned a lot from all of this. Hard times teach us so much, I worry that all of the "modern conveniences" are robbing our youth of these lessons, perhaps. I often went wrong and had to "fix it." For that I'm not bitter today—in fact I'm actually grateful. I am stronger now. I had to use my brains to develop my intelligence. And what I lived through made me understand a lot of different situations. People who are out of control themselves often try hard to control others, and this doesn't work well for them at all. You have to be an empowering person. Otherwise this does not work for you at all because people are the most important asset you ever have. Without willing people, you do not get anything done at all. So the treatment of others is very important. Give positive reinforcements always, be a good cheerleader. But I just didn't understand that back then. In fact, for years I thought I was crazy. I would try to please everyone. We often have to *learn* how to stop blaming ourselves, learn to stop the self-hate. We are not responsible for others' emotions, only our own.

After a while I wasn't a nice person anymore. I hated myself. I didn't want to hurt or kill anyone, and that's what I was afraid of. I knew I was capable of it. I had a quick, sometimes uncontrollable temper and I used to zone out. Not good! I just didn't like myself anymore. My self-esteem was very low. The only time I felt good was when I was drinking. I got into drugs a little, but I didn't really care for pot because it would put me to sleep and I would get paranoid and itchy. I liked the alcohol better.

A lot of guys I knew used to work for beekeepers in California, so I called a beekeeper there, got a job, and away I went. I got on a bus and travelled to Yuba City, California. After getting a work permit, I worked there for about two months packing bees (Canada used to buy all their bees from California). I was making two dollars an hour and I loved it. And after the packaging season was over I got another job for another beekeeper, again in California, and worked there for the summer with some guys I knew.

I had the time of my life in California, with the warm weather, and all when I was 20 years old. We did a lot of partying and we had a lot of fun. We used to go swimming almost every day, all meeting at these swimming holes on Stony Creek in Bidwell Park in Chico, California. Really good times. Some drinking and some pot, but not out of control. There were a lot of nice girls around, and their boyfriends were all in Vietnam. That's all I'll say about that, but we did a lot of sitting around the swimming hole and it was just a great time. We played ball with a church team and had a great time. A man we knew had a boat and we used to go water skiing. My friend Bob and I spent a lot of time at his house. He was married with two young children and he would let me stay over, his wife Pat being such a nice lady.

I met a young man named Roger and we worked together in the honey bee business. One thing led to another and he invited me to his house on a Sunday for dinner, to meet his dad and his brothers and older sister, Marie. She was a very pretty lady and we ended up going out together. We had a lot of fun, swimming, partying, and just having a good time. I got to be very good friends with the whole family. After a while they started saying I was their adopted son from Canada—in fact they still call me "brother" and their kid calls me "Uncle Paul." I love them a lot. Hubert became my new "Dad," and Margaret my new "Mom," giving me a new family. Unfortunately, as time passed, when Marie and I would go out she started being more interested in other boys. I eventually had enough of this and we ended it, but we all stayed very good friends.

I've visited them often ever since, but Hubert passed away four years ago. I traveled to the funeral from Canada to California. It was hard, but I really did feel like part of the family, and I wouldn't have missed it. I'll always be glad they took me in back then. It helped so much, being part of the family. Thank you, Trisha, you're a very nice lady. Roger, Marie, and Clinton—the whole family is just great. Wendell and Brandy, Thomas and family, I know I haven't come to see you lately, and Weston and Elisa, someday I will show up, I promise. The Lower family loved me the way I was. I drank a lot back then, but they never gave me a hard time at all. They were always loving.

"Paul," they'd say, "someday you're going have to look at your drinking. You'll have to figure it out."

Mrs. Lower's mother had quit drinking a long-time prior, and she had terminal cancer. I spent a lot of time talking to her, and she would tell me I was a nice young man and I was going to be okay. She was always positive and very nice towards me, and I didn't realize it back then, but this helped me so much. They became a part of my life, and this will be with me forever. I am very grateful for their kindness. They showed me total acceptance; they made me see I wasn't the awful person I believed I was.

How much should we love each other? How often should we speak and stay in touch, should we visit? A lot of families don't visit at all nor do they even talk much, and there are reasons for this. Some family members think they should control or influence each other, but this does not work very well at all. In fact, you're going to make enemies this way. As families, we need to accept each other exactly as we are. We should help each other if we can. But trying to change each other does not work at all. The Lower family accepted me the way I was, and it worked. Love each other the way you are. You cannot change each other, but you can change yourself and accept your family members the way they are. Living with family and knowing each other as friends are totally different. Treat your family like your best friends. Total acceptance is the key with family. Accept the things you cannot change, change the things you can, and pray for the wisdom to know the difference.

Before I left Canada for California I had been charged with "illegal possession"—meaning I was too young to have alcohol—and I had received a speeding ticket. I left for California without paying these tickets, so law enforcement was looking for me, with a warrant out for my arrest. Well, they caught up to me and arrested me. I had around $70 or $80 on me at the time, and this was a good amount of money back then. They took all the money I had, and I had nowhere to go, so they took me to a homeless shelter. It wasn't a nice place at all—it was old and full of homeless people. And all these poor guys—I didn't know what to make of it. But let me tell you, I never ever wanted to go back there again!

The next day I ran into an old friend named Cassidy at a service station. He was a truck driver and I had worked with him south of Grande Prairie, operating a loader and loading logs. I told him what had happened to me, and he said he was going south to Fox Creek and could give me a ride. He also loaned me enough money, so I could eat. I was very grateful because he didn't treat me like a low-life, even though I felt like one. I think back, and these are all people that made a big difference in my life. These are the people that become lifelong friends. Years later, it was Cassidy who got me my powerline job, another life miracle.

But the biggest lesson I learned was that I knew I never wanted to live like that again. I saw these people, in the homeless shelter, and I don't know but I think they are powerless over their own lives, broken. And I never wanted to be broken or powerless again. And if you aren't able to live on your own, you really are powerless and broken. I had nowhere to go and nothing to eat. While it wasn't for long it was certainly long enough for me to get the message, and you know what? I was never broke—not ever again. That time and experience had a *huge* impact on me (again, learning from hard times) and it has stayed with me to this day. Things happen for a reason, and they happen in mysterious ways.

[6] TO BE A LINEMAN

I WENT BACK HOME to Canada, and I worked in a logging camp south of Grande Prairie. I worked at Camp Five for Canfor Lumber operating a log skidder, falling trees, and I was doing very well at it. I worked for about five months before the spring breakup (we all used to get laid off for spring break). I know a lot of people in Grande Prairie, Alberta quite well, and on this break, I met a guy named George White (I will always remember his name) who worked for Alberta Power. My friend Cassidy introduced us.

"Paul is a good machine operator," Cassidy told George. "Ask him if he will operate a crane for you."

I told them I had never operated a crane before, so George asked what I *could* operate, and I told him Caterpillar trucks and log trucks, that I had operated log loaders and a gravel loader as well. He asked me if I was from a farm and I said yes.

"You're not going to have a hard time figuring this out, then," he said, and he introduced me to Wally Gilbrey, the supervisor for ACE Construction out of Calgary.

Wally asked me questions about what I had done, and he then gave me a job.

"Where do we stay?" I asked.

"In a hotel," he said.

"I can't afford to stay in a hotel," I said, and he assured me they would pay for my room. *Wow,* I thought, *they pay for your room and they gave you five dollars a day for meals!*

"As a crane operator, we'll pay you $3.25 per hour and overtime after 48 hours," he said.

I wondered if such a thing was possible! *Am I dreaming?* I was in Heaven. *Thank you, Cassidy! You just introduced me to my new dream job.*

I went to work that first day, not quite knowing what to expect. The head supervisor running the project was a guy named Ernie Skavlabo, and he was quite a guy. He looked old to me, like about a 50-year-old lineman. I started on a Cat (Caterpillar tractor), stringing wire. I didn't know what do at first, but I picked it up pretty quickly. We were stringing a long line, all the way from Grande Cache Alberta to Grande Prairie Alberta, around 80 miles or so. I hit it off with these guys, too. They were pleasant to work with and they all worked hard. And I was really enjoying operating the Cats and the cranes. I also drove a line truck and I hauled fuel for all the equipment. And we worked long hours. The line we were working back then was a 144,000-volt (144KV, or 144 kilo-volt) transmission line, a good size in those days. Nowadays 500,000 volts is the norm, but back then the highest voltage was a 240,000-volt transmission line. Times have changed so much. But I liked those guys. We joked around and laughed a lot. They were a fun bunch and I felt like I fit right in. I loved every minute of it, to tell you the truth, and I had found my star.

One day we were "sagging conductor," or tensioning the conductor line. "How would you like to climb a pole?" one of the linemen asked me. "Why don't you give it a try?"

"Okay!" I said. I had been asking a lot of questions. "I don't have anything to lose, right?"

"No, you don't!" he said. "We can just go up around 20 feet, and if you don't like it you don't have to try it again."

I was a little afraid, but I went for it. They got me another lineman's climbing gear with the belt and spur, and they helped me put them on the proper way. He belted under me and we started going up, and that wasn't as easy as it looked at all, let me tell you! As you're climbing the pole, you're balancing on one leg because you're using the lineman spurs/gaff[7]—and you push yourself straight up with one leg! Not easy at all. I was about 190 pounds then, and I was in good physical condition, but I had not realized that balancing off a small gaff takes a lot of strength. I didn't realize it then, but I was about to change my life forever. We went up maybe 10 feet or so and he said, "Let's go back down." Now, it's actually even harder to come down because the balance you need is unbelievable going down. The second time up we went about 20 feet high and we had to get work done up there, so that was all the playing we could do that day.

I felt different. I wasn't sure about climbing poles, but I was excited somehow and it made me think. It was a challenge. It was different. I asked myself, *Do I really want to be a lineman?* This was the first time I really thought about anything like that. All the other work I had done was okay—I did it because I

-----

[7] "a climbing iron or its steel point used by a telephone lineman" — www.merriam-webster.com/dictionary/gaff, as strapped to the lower legs with a hook or spur for climbing

needed the job and the money for survival—but I had never considered anything prior as a career. The next time I went up the pole I climbed all the way to the top. Once there, I was shaking, nervous, excited, pumped up—and I felt really good. I told myself, *I did it!* It was hard, like my first horseback ride, but I had achieved something different and I felt like I was on top of the world. I wanted to climb more. I knew this was what I wanted to do, to become a lineman.

I was hooked.

My dream was to become a power lineman.

I started climbing as much as possible and I still operated equipment. After the job was completed I traveled to Provost Alberta to work on a "138.000 KV"—138 kilovolts line. I worked on a setting crew, setting structures using a conventional crane and pike pole. We worked hard shoveling dirt and tamping poles. It was a little different this time because I was working with different guys, who I didn't know. I missed all the guys I had worked with before, but in line work, that's the way it is. This is why they say don't make anyone angry because someday one of these guys might be your foreman, so be nice to each other. Linemen are like family. They are a very close-knit trade, and this why I liked it so much.

We all moved to Rocky Mountain House to work on a 138KV tower line. I went work on an assembly crew bolting steel with all different guys again, and I learned how to read blue prints. It was good work. We assembled all the towers, then I went to work on the erection crew setting towers with a hydraulic crane. I was working as an apprentice lineman now. After we finished the tower line I operated equipment and again did line work. And when the job was finished I went to work on a job in High Prairie. We worked on a 25KV distribution oil field line, which was 50 miles long. I was on the string wire crew with a small John Deer Cat. I climbed a lot on this job, and I operated equipment. The foreman was a good guy to work for, named Isidor Hurarusha, a good old lineman with a lot of hands-on experience. He had started, I think, at 16 years old in line work.

One day we strung 70 structures and tied in 68 structures. In those days, we were going flat out. The foreman would drive you from pole to pole, and we each had a hand line (we pulled the wire up by hand). The Cat, the trailer, would back up so we had slack in the wire and we would pull it up by hand, with a ground man (those who did all the ground work) on the ground. We climbed up the pole as fast as possible, hung hand line (the rope we pulled the wire up with) on the end of the cross arm (what the insulators fit on atop the pole), and we would wait until the wire was past the pole about 125 feet or so, then tie the wire onto the hand line. You then started pulling the wire up. The Cat would back up and the wire was tied with half hitches (knots that remain when you pull up but unravel when you pull down). The man on the ground would let go of the line a bit, the half hitch would come undone, and you would grab it from the top of the pole, and you just held on to the wire until it was over the cross arm, at which point the wire was secure.

It was very simple, actually, but it was fast. The wire was set for the weather. We determined how much sag to leave in a segment of wire by the tension and the temperature—the colder the weather, the tighter the line. And the crew behind us tied in the conductors.

We worked hard, and we played hard. We drank beer, and it was a great time. I really enjoyed my new job—loved it, actually. I caught on fast, too. In line work you have to visualize and remember the job procedures. I could be shown a job task once and I had it. I remembered most everything. In line work you have to be good at mobilizing and demobilizing equipment because you're using trucks, Caterpillar tractors, cranes, setting up reel trailers, and you're always working off-road in ditches and in very rough terrain.

Then we traveled to Swan Hills, Alberta (a big oil field area), building oil field line, three-phase (for larger, industrial applications) at 25 KV (25,000 kilo-volts). We hung a lot of 30 KVA (kilo-volt amp) transformers (which stepped-down the voltage to 480 volts). I learned a lot. The equipment we used was not very modern compared to what we have today, but that's all we had, things like gin pole trucks and highway diggers. We didn't have a lot of tools. We used rope blocks (heavy rope, 5/8" or 9/16" rope that will pull about 600 pounds), and bull ropes (about an inch-thick) to hang larger transformers. We had no drills at all, we used a brace and bit[8] (or "sawdust pump,") for drilling holes in the poles by hand. Up there on a pole they are hard to use. You had to balance on one spur while you did it, from 40 to 60, sometimes even 110 feet high. You had to work smart, and it was hard work.

I don't think working hard makes you smarter, but with a hard job task you usually have to work smart to be successful and stay safe, and by so doing you develop practical experience and job skills. These older line hands were smart, they were practical, they clearly had *skill*. Those guys could rig anything up. I remember for a long time we didn't have chainsaws. We used cross-cut saws. You learned to use hand tools, and you learned the importance of maintaining tools and equipment very carefully, so they would function properly. Otherwise they become very useless. Ernie, a fellow lineman, said to me one time, "You have to outsmart the shovel." Some people are surprised to hear there is a right and a wrong way to use a simple hand shovel, but there is! Take back-filling holes by hand, for example: You put the shovel on your knees; put one hand low, near the blade, and then sit down, pushing the shovel

---

[8] A **brace** is a hand tool used with a **bit** (drill **bit** or auger) to drill holes, usually in wood. Pressure is applied to the top and the tool is rotated with a U-shaped grip. —Wikipedia

forward and the dirt into the hole at the same time. And it works!

[7] WORK PERILS

FTER A WHILE, I went to work in Hay River[9]. I had not done any work yet on 4160–the old primary voltage system. In fact, I had never done any rubber glove work, and didn't know anything about it. I learned from older linemen to cover the other phases and cover neutrals, grounds, and guy wire[10] when working hot line (working on

---

[9] "Hay River, known as "the Hub of the North," is a town in the Northwest Territories, Canada, located on the south shore of Great Slave Lake, at the mouth of the Hay River." —Wikipedia

[10] "A guy-wire, guy-line, guy-rope or guide-wire, also known as simply a guy, is a tensioned cable designed to add stability to a free-standing structure. They are used commonly in ship masts, radio masts, wind

power lines while they're hot/electrified). You have to always eliminate the second point of contact, as it's very dangerous work. We had to hang 14.4 transformers on the same pole as the old transformer, and if there wasn't enough room we hung the transformer one span down. The older line hands talked about how they used to use leather gloves, but we now used rubber gloves to insulate you, while keeping your feet clear of grounds and neutrals (you complete the circuit that way and can end up dead). That's how you worked with high voltage.

With lower voltage, you can get away with touching a primary if the pole you're working is dry and you're not also touching grounds or guy wires. But it's not a good or safe work practice at all. We didn't try it but these older guys talked about it—they killed a lot of linemen in the old days. I guess in the early 1900s about *one out of every three* linemen got electrocuted and died. Hard to imagine nowadays that many men dying on a job other than in war. A man named Harry Miller soon enough started the IBEW (International Brotherhood of Electrical Workers) union in the name of safety. And it really wasn't about money or benefits, it was all about safety. Even now safety is the most disregarded aspect of the job, because so many simply think working safely is going to slow the job

---

turbines, utility poles, fire service extension ladders used in church raises and tents." —Wikipedia

down. Well, when I speak or train, I ask, "Wealth or health, what's more important?" I, for one, say health. Life is about making wise decisions, and safety is a good decision. I like to say that safety is also a good attitude.

But today it is a little different, today we have a higher regard for safety and being in control, when possible. There are posters in the workplace touting safety. Companies bring me in to talk about it nowadays. Take driving a car, as an example. How do you eliminate the danger when you're driving your car down the road? Well, you can only usually control the hazards. If there's a person driving that has been drinking, how do you control them? You don't! All you can do is be a defensive driver, be proactive rather than reactive. And it's like this with a lot of different work tasks. It was very dangerous in the old days and it's still dangerous nowadays, if not quite as bad because we work out of bucket trucks and we have insulated buckets. In the old days, we worked off the pole all the time. We didn't have bucket trucks at all, so we always had a second point of contact with our bodies while working with plenty of voltage. There is also a lot more hazard assessment training today, so we don't expect to injure workers on the job now like they used to.

Everything was done on the job in those days. For doing primary work we had our hot line hoses and hoods and rubber blankets insulated at 5,000 volts. For sagging (pulling) wire we used rope blocks (thick, insulated rope) for everything because it was insulated. Definitely the old way of doing things. We had few, precious tools, but we had what we needed. We worked carefully, with practical skills, and there really was no safety at all. And we didn't do any safety training at all, either. They call them the "good old days" (my father used to call them the "good olds") but they were really not good at all! My father said they worked like dogs and they were treated like dogs. In my early days, we worked hard, drank hard, and played hard, but I wouldn't trade it for anything. Being a lineman was my calling, and I really enjoyed it a lot.

I traveled for work to Rainbow and Zama Lakes[11], and worked in the oil fields under a foreman who was usually drunk, building three-phase lines to oil pump jacks. Alcohol had basically taken over his life and he was not pleasant to work with, with lots of emotions and drama. Your personality affects those who work for you, good or bad. If you dislike yourself, you tend to dislike others. That's important to

---

[11] "High Level is a town in Northern Alberta, Canada. . . The area surrounding High Level is known for its oil reserves and forests. Two large oil and gas fields, Rainbow Lake and Zama, are located west of the town, which provides services to the oil patch." —Wikipedia

understand when you encounter someone who mysteriously dislikes you. It's often nothing to do with you at all.

It was so cold there it was unbelievable, and I mean it was like in the *negative* 50s and lower—so cold we were often not going to work at all. This was in December and we had to have those oil field lines built before we could go home for Christmas, so we decided to work in the afternoons. We would get the trucks started and let them "warm up" (to about negative 40 and 45 degrees (Fahrenheit) and we then started setting poles with an old highway digger. It was a good machine and we could set from 20 to 30 poles each day. It was so cold we walked from pole to pole to stay warm. We found if we let the truck run all night we didn't have to struggle to start it in the cold morning because the truck's block heater wasn't enough, and unless we left it running, we had to use a tiger torch[12] to help get it going.

It's unbelievable how you get used to the cold, tougher than hell, but after a while we actually did get used to it. We started working around 9 a.m. or so, because again, we wanted to get done so we could all go home for Christmas. We were only a four-man crew and the alcoholic foreman wasn't helping much. In fact, he usually didn't even get out of the truck. We

---

[12] a type of propane torch with a wand and extension line

worked hard to get finished. As we could start to see the end of the job, we just started going flat out. And we finally had set all the poles. I can't remember how many oil field taps we built, but we worked for over a solid month. I think we strung and set 30 miles or so, I can't remember the exact amount. And it started getting even colder, like into the negative 60s, but we wanted to get the job done before Christmas, so we could go home, and I wanted to see my mom and dad.

We were happy to get finished. At the end of the job we traveled to High Level Alberta[13] where we partied and had fun. The useless foreman started mouthing off and got a "tune up," got what was coming to him. That was how we took care of "anger management" back then. And I didn't want to work for him anymore. He left the next day and I did see him again about four years later. He still had a bad drinking problem, and he ended up drinking himself to death, actually, unfortunately. I look back and think he was a hell of a lineman, but alcohol had taken over—to be so miserable and *think* you're so happy. Alcoholism is a sickness and it cannot be healed. Once you have an addiction all you can do is stop using/drinking, and

---

[13] "High Level is a town in Northern Alberta, Canada. It is located at the intersection of the Mackenzie Highway and Highway 58, approximately 733 kilometres north of Edmonton and 725 kilometres south of Yellowknife, Northwest Territories." —Wikipedia

learn how to live without it, to stay away from it and clean yourself up.

In Alberta, the line contractor used to shut down for two weeks at Christmas, so this was when everyone took vacation and all the equipment went to shops for maintenance and repairs, and a *lot* of that was done in those two weeks! I traveled to Vancouver Island[14], as my parents lived in Nanaimo, British Columbia. It was nice. I took the train from Calgary to Vancouver, and this was a lot of fun. I remember the price of a one-way ticket was just 38 dollars. On the train, just before Christmas, there were a lot of people in a party mood, and well before we got to Vancouver we drank the train dry. It was a two-day trip and we partied all around the clock. The bartender said, finally, "You guys drank us dry! No more alcohol!" But we had fun! I still don't know if he meant that or if it was a nice way to cut us off.

When I got to Vancouver I must say I was pretty burned out. All I wanted to do was sleep. I took the bus to north Vancouver to the ferry terminal and got on the ferry to travel

---

[14] "Vancouver Island, off Canada's Pacific Coast, is known for its mild climate and thriving arts community. On its southern tip is Victoria, British Columbia's capital, and its boat-lined Inner Harbour, neo-baroque Parliament Buildings, grand Fairmont Empress Hotel and English-style gardens. Harbour city Nanaimo, home of chocolate-and-custard Nanaimo bars, has an Old City Quarter with shops, galleries and restaurants." — Google

to Nanaimo. I had seen the ocean once before and I wanted to sleep, but I found myself awake, watching the scenery as we went. It was beautiful country—the mountains, the trees, the blue ocean—I just loved it.

It was great to see my mom and dad because I had not seen them in over three years. Even seeing my dad was okay. I had been there for about ten days or so when I was looking in the paper and saw that a company was looking for linemen. I called, and it looked promising. I met with a man named John and he hired me, so I traveled back to Calgary, got all my line tools, belt, and spars. I then traveled back to Vancouver Island and went to my parent's house, and when I arrived, my dad told me he didn't want me living there anymore. That hurt some, but it was okay, because I preferred not staying. I knew what my dad was all about and didn't really want to be a part of our family anymore. I had had enough, and I had learned how to survive on my own and liked my freedom.

That was the story of my life up to that point—rejection by my dad. But this time his rejection didn't hurt anymore, I had become immune to it, and I knew I could make it on my own. Call it being cold, if you wish, but we all do what we have to do to survive. And I just didn't have any desire for that treatment anymore, and that's actually a state of mind that can become dangerous. It's why you do not want to pick an argument with someone who has been abused. It can turn not in your favor. It's a feeling that's hard to describe, but one that goes very deep. And I just didn't have any feeling at all for my dad anymore. I would look at him and see nothing at all. If I would have never seen him again, it wouldn't have bothered me at all. I went into downtown Nanaimo and found an inexpensive hotel room I rented by the week and stayed for two months. And I wasn't angry with the situation. Acceptance is powerful. It came from within and I was at peace with myself.

Some make bad decisions because they make them based on emotions. My father made a lot of his decisions that way, out of anger. It's a matter of self-will. Anger blocks your mind from making good decisions, whereas a state of calm yields good decisions, normally. Anger is very reactive. I didn't understand that in those days, so I responded with self-doubts, with resentment, anger, and sometimes fear of people hurting me. I learned to react and push back, and this doesn't work very well in life. Dad tried to empower himself with anger and he used it to control others, so I started staying away. People like that end up as loners. You ultimately just avoid these people to stay out of arguments and it only works for a short time. All you do is push back and forth, no one gets ahead, and it just gets worse and worse. I wish I would have known then these skills that I know now. These people are so trapped by their own unhappiness they become very self-centered, and you come to realize that positive people discuss the future, they discuss productive things, so you choose who to spend your time with, and you either suffer or benefit from those choices.

I was starting to accept what my father kept telling me, that I was trouble and going to end up in jail, and that I was no good for anything. But I had worked for powerline contractors and they liked me, they liked my work, so my confidence was coming back. My bosses couldn't believe how fast I could pick up powerline procedures given I had such a hard time in school. Ernie, an employer, told me, "Paul, you are not a stupid guy! It's amazing how fast you pick this up. You can put a lot of education in a person's brain, but you can't put intelligence in a person's brain. Some people have good athletic skills, some have very good book memories, but you have good practical skills and good hand coordination." And it was true, I was always good in sports and I was a very physically strong person. I could learn reasonably well, I just had to just go about it slower than others (so did Einstein, by the way!).

I started working for a small line contractor (some of these people may still be around), and they were good to me. They would say I was a hard worker, that I wouldn't stop for anything at all. My father used to say, "Never think you are smarter than the boss," and back then this was the way things were run. We had a sheep mentality, to do as you're told. As a result, we did not ask questions. We didn't want to get ourselves in trouble. You never questioned those above you. That's how it was for the working class back then, it was a "pee on" mentality.

I started working with an older line hand who was a good lineman named Ivan Wallace—actually, he was a great lineman. We started doing voltage conversions, upping the voltage from 7,200 volts to 14,400 volts. We had to change all the insulators with hotline tools, like insulated sticks, working off the pole. I didn't have much experience, but I learned quickly. They could even treat me like shit and I would take it, I was so used to it. I just wanted to do well for myself. My dad had in fact made me that way, to focus on the job and really make good on each job task. We hung dual-voltage transformers, and it was good work. Old Ivan knew his hot line work and I enjoyed working with him and learning. I just loved line work and Ivan was a mentor of mine. He could see a lot stuff in me, including the problems. I was a people-pleaser, I just didn't want to get into trouble. And that wasn't good because I would take the abuse until I would implode, and then it wasn't nice at all.

Back then, I would let people pick on me and I wouldn't say anything at all, wishing they would go away. Ivan saw and was very helpful with this. He wouldn't let others pick on me because he had been through the same ordeals. He was kind to me, and he helped me a lot. He could see the anger in me grow and he knew that if I lost it, it was crazy. I had no control over my temper. I did some not-so-nice things to people who were not nice to me, and I didn't know how to handle those things any other way. I didn't even warn those who picked on me that I had had enough, and to just back off and stay out of my way. To this day I have to be very careful with my temper because I can lose it very easily if I let people treat me like I used to be treated. I can still, sometimes, go right back into "combat zone" where the adrenaline kicks in. I get so hyper about it that I get a sore back after I calm down. So I still have to be very careful with these situations, I still have to get out of the defiance mode, and it took some time just to get to where I am now, aware of all of this.

[8] THE BEST ADVICE

I MENTIONED ERNIE SKAVLABO, an old, Norwegian lineman and foreman, a big guy—my first foreman, in fact. I learned a lot from Ernie. One time, the other linemen and I were waiting for a supply of wire to arrive. Ernie showed up and asked what we were doing. "Waiting for the wire to come in," we said.

"There's always something to do," he said. "There are tools to maintain, trucks to clean. I never want to see you do doing *nothing*. Who do you work for, anyway?"

"Ace," I said.

"Wrong," he said. "You work for *yourself.* Your name, your reputation, your integrity, your honesty, basically it's all you have. Your name and reputation are always on the line. They're all you have to fall back on, your name and reputation. They're really all you have."

Well, that has stuck with me forever. What do you *have* when you leave a job or a company? To this day I have always considered I work for myself, and I have a clear conscience. You may think it not important, but at the end of each day this is all you have. It's a good feeling and it has saved me many times over the years. It has kept me going in not-so-nice times at work. It has helped me separate myself from ordeals at work, when other workers get into those negative controls they slip into. They get very angry with the company they're working for, and they get sidetracked *big time* and the end result is never good. It's the best advice I have ever gotten, and ever since, I have gotten many letters of recommendation, just by doing my job. A day's work for a day's pay is all I did. The company has to make money, too, to keep you employed.

I continued working voltage conversions and re-conductoring, setting poles in hot lines, and I just loved it. We finished the voltage conversions by Nanaimo and Cedar Gabriela Island and we had a great time. I loved Vancouver

Island and worked some other islands around Campbell River such as Tuxada Island, and Salt Spring Island. We built a lot of single-phases to new homes and three-phase 25 KV lines for commercial use. On Vancouver Island, we had to do a lot of rock drilling and blasting, and in all different situations. Very demanding work. It was my dream job and I loved it, loved every minute of it, and if you like your work, it's not really a job.

I traveled a lot with my job. The company we worked for in British Columbia had a rule that if the job was a government contract you had to be a union member, and the owner of the company didn't like this at all. He wanted us to join a smaller union, but all the workers wanted to join the IBEW (International Brotherhood of Electrical Workers) because if you belonged to the IBEW you could travel all of North America. After confusion and disagreements, I ended up working back in Alberta, with no work and no union, so I went back to work in the oil fields, building three-phase oil field lines, and it was good. I learned that I loved Vancouver Island and I missed it a lot, but I stayed in the oil fields for about two months. There was a lineman from eastern Canada there and we got to be good friends. We talked about working back east. He told me of a line company, so I called and they told me they would put me to work, and away I went to Cornwall, Ontario.

I worked there for the winter for a company called Eastern Pole Lines. It was all high-voltage rubber-glove work, changing poles for public utility companies. They had a big substation rebuild to do in Cornwall and I stayed until spring. By then I was getting pretty good at line work and I loved it.

In the spring of 1975 I traveled back to Alberta and went to see the guys at Belcourt Construction in Edmonton about a job. In those days, they just asked questions about different jobs you had done, and they knew by your answers if you knew what you were doing. I told them about the hot line work I had done, so they put me to work, and I was happy to be working on a hot line crew. I learned a lot. We used to change a lot of hot air break switches because they wear out over time.

That spring there was a big ice storm across all of southern Alberta. The power was off from Red Deer south to Lethbridge Alberta, and we worked for two weeks straight to get it fixed. We would start at 6 a.m. in the morning and work until midnight, sleep for four or five hours, get back up, and just keep going. This is what made me think, *Do I really want to stay working as a lineman?* Well, I did, and I still loved it. They say once you're a lineman you will always be a lineman because it's in your blood, and I believe that, even to this day.

I ended up working in Jasper, Alberta, converting 4160-line to 25KV, doing a lot of hot work, rubber-glove work, and ended up working for Ace Construction on the hot stick line crew. I traveled all over, changing burned-out insulator switch poles. It was good work, just like the work I did with Belcourt Construction, all hot line work, and again I loved it. I worked with two nice guys, Percy and Hank, who both had good personalities for doing line work. With my recent experience, I knew a lot more about line work, and it was a lot easier for me at that point. I was now figuring things out on my own.

I worked for different companies, I worked all over. I loved traveling, and when a lineman works all over like that, they call him a "tramp lineman," which is no slight, it's accepted. And that was what I wanted, anyway, to travel a lot, and to this day I still love traveling. A coworker once said, "One thing about Paul, if he is here he just came from somewhere, and he is planning to go somewhere else he hasn't been." I just loved the travel, loved doing different projects— in fact I wanted do it all: distribution line work, transmission line work, and underground line work. I worked in Mika Creek BC, Trail BC, Christina Lake BC, Vancouver Island, the Hudson Hope Dam, and more. I worked in Fort McMurray in the oil sand for four years, and then went back to Vancouver Island. And that's when all hell broke loose.

[9] HEAVEN & HELL

WHILE DOING ALL OF THIS hot line work, I had also been drinking a lot. I was just having a good time, but not knowing the effects of alcohol. It's a depressant and I got into a lot of unnecessary stuff because of the drinking. I got into a lot of fights. My friends started staying away from me and they didn't want to have anything to do with me. I lost very good relations. It was bad. I lost a woman I cared for, a very nice person who did not deserve that treatment at all. I never was physical with her, she just had enough of my drinking. She ended the relationship and it hurt, and that was when I really started looking at myself.

One night I got into an altercation with two guys. I was drinking, they came after me, and I blackout-imploded. When I came around and became aware again, one guy was on the ground, dragging the other guy to the sidewalk. I wanted to break his knees. My friend started screaming at me to stop, which I did, and we took off as fast as we could. That was my wake-up call. I was so afraid that I might not be able to stop drinking that I couldn't sleep anymore. And that's when I told my friend I was quitting drinking.

I was working in Port Alberni, BC, and the union wanted to talk to me, and I had no choice. I was told to go and speak with Jack Zetller, who I had met previously at the union hall. I sat down, I knew what was coming, and he closed the door. I always liked Jack because he was a straight shooter. He asked me if I had any idea why I was called into the office.

"I think I do," I said.

"Paul," he said, "we are not saying you're not a good lineman, but your drinking is getting in the way of your work. Paul if you would quit drinking you would be a *great* lineman. And you would work your way up in no time. If you don't quit drinking, I am not going to give you any more work clearances."

I had no choice at that point, I had to quit, and I did quit. I went home, and I had my last few drinks with my friend Dave Bartee.

"This is it," I told Dave, "I am quitting drinking."

Those were my *last two drinks* and I have been clean and sober now for over 38 years. My brain is doing a lot better, and I am a happier person now. I did get help from a 12-step program, which was what I needed. And it was hard. When you drink you bury your emotions, you never actually deal with problems, and the brain damage you do to yourself takes time to rebuild because alcohol and drugs are depressants, after all. When I was drinking I hurt a lot of people, and I had to make amends. I had to clean the past up, and this wasn't easy at all. I didn't go to a detox program, I did it on my own, but if I could go back and do things over, I would go for rehab. Again, it was not easy. I went through withdrawal. I drank for 15 years and they say it stunts your growth (all kinds) from the time you start drinking. Luckily, I had a lot of friends who loved me and cared for me until I could love and carry myself again. In hard times, true friends will help you. They will never stop loving you. I really found out who my true friends were. My drinking buddies never showed up to help me at all. One true friend, Jimmy, has been my friend for 35 years now. To this day we call each other brothers.

The biggest problem though, was that I did not *like* myself at all. I remember a lady told me to "stand in front of a mirror and look at yourself and tell yourself you love yourself." Well, I tried that, and I just couldn't do it. It was just too hard. It took me a long time but gradually I did it, and it's so funny because once I did it, I started laughing, had the best laugh ever up to that point, I believe. You have to love *yourself.* There were battles inside that I had to deal with and I did. And the funny part is that bad things had happened to me, and I was blaming myself for them for the longest time, but now I can see the difference. I realized how much I hated myself. At one time, I was so depressed I didn't want to live anymore. I just could not forgive myself for being so self-centered. I just hated myself. I had lost everything, and I am not now talking about money, but my self-esteem. It's very hard to be positive when you're in that bad a state of mind. The longer I stayed off the alcohol, the better I began to feel, understanding that alcohol will induce brain damage (15 years of it). When we take depressants to make ourselves happy, it doesn't work.

It took at least three months for me to stop craving alcohol, and in that time, I was an emotional roller coaster with big mood swings. I would get angry easily. It was a matter of learning how to cope with all of those emotions as I had drowned them out. I didn't know how to handle certain situations—if someone took a shot at me, I would take a shot right back at them. Well, that's not the right thing to do at all. I had to learn to not react to every situation, and this took a long time. I had to learn how to get away from the wrong kind of people. It's like my old friend Monica told me, "Look for the smiling people, stay away from the angry ones."

I had to learn to put up with negative, manipulative people, to learn how to think, how to make decisions while keeping your emotions out of it. I had to learn new coping skills, how to not always react to emotions, how to stay focused on my daily tasks. If someone gives you a negative push, don't fall for it. And the funniest thing is that I see people who are compulsive liars who are very smart, and this is what they do (they think) to survive. They start at a very young age and this is all they know, that's just the way it is. But all you have to do is let them hang themselves. All you have to do is wait and they will destroy themselves. At work or in life, I stay out of these conversations. Refocus on yourself. After all, why are they attacking you? It empowers them (they think). They're insecure and they just want to control your thinking. I had to be honest with myself before I could be honest with other people, and being honest with yourself, believe me, is a lot more difficult.

In today's world, there are a lot of emotional subjects like politics and religion. Media propaganda is very powerful. We have to learn how to think in a way that sorts truth from lies. We have to be confident with our own reality. True thoughts *are* good thoughts. Good and honest observation leads to good decisions, as long as we can separate out our emotions. Look only at principle, not personality. When people fail in some way they tend to cover their own butts first, so this is why I choose to tell the truth the *first* time. It's a practice that will save not only your own butt, but your friend's as well. If you lie, it usually catches up to you, so the reality is you do these things to yourself. People lie out of fear and intimidation.

When you do find honest people, keep them as friends forever. Honesty is hard to find, trust is hard to find. Honesty and trust in yourself are priceless. If we need help and we don't get help we will stay weak. Emotions take over. The way I survived and got stronger was that I finally admitted I needed help. I started winning. The best decision I ever made was on the day I decided to quit drinking. I admitted defeat, but I conquered my weakness in the same instant. I was the winner that day, and my life has changed completely. I immediately started living a new life.

It's funny, or perhaps not funny, that when we have problems, often the first thing we do is go to alcohol or drugs.

This is, of course, not a good mixture with emotions. Alcohol and drugs only mask or hide the problem, and the emotions get out of control: "self will run riot," as they say. The biggest emotion is fear, which leads to anger and resentment—all bad emotions. These lead to bad decisions. We can all fake being all right. We can be just doing terribly and tell ourselves we are doing just fine. Don't lie to yourself! Self-honesty is the only way you will preserve your integrity. You don't have to admit everything to the world, you only have to admit things to yourself. It's the road to feeling good about yourself, too. Honesty starts within yourself.

When people try to use emotions on you in order to control you, it's actually very easy to recognize, and the solution is simply to make that sure you do not react emotionally. You can also ask them questions like, "What's going on? Is everything okay? Is there anything wrong?" Most of the time they calm down. Try and get them to refocus on their own thoughts by asking questions like these, and not in any kind of sarcastic way. They have battles going on inside themselves, after all.

I had to *learn* how to communicate in a positive way. I had to learn to just do what I had to do and stay out of the "humorous" spiking, the making fun of anyone who made a mistake. This goes on more than you think, in all kinds of jobs and circumstances. Maybe you can think a of a few. I had to learn how to be helpful with these individuals. I am still working on this, and I probably will for the rest of my life. When you see someone taking cheap shots, it makes you wonder why. Most likely their parents used these negative controls on *them,* and it's all they know. But remember, they take these shots out of fear. Their parents didn't show them how to communicate, and I wasn't shown how by mine, either.

I had to learn on my own, and stopping alcohol was what really got me thinking because I had nothing to camouflage my inner issues anymore. Staying clean helps you clear your head and you start to produce serotonin again, our natural "happy juice." It's easy to say we should all be more positive with others, but how? I myself just didn't know because I had none of these skills at all. My mind was all over the place. I just didn't have the coping tools I needed. *Others* used to control my thoughts, and I always wanted to please everyone, so it was all reactions, no people skills. If someone challenged me in any way, I would take it personal and I would fire up easily. I had to learn to stay out of the personalities and refocus on the task(s) at hand, because there are a lot of different personalities out there, and the only one I can control is my own.

By each doing our part, whatever that might be, we can become good team players, and in turn become good team leaders. Life then becomes a lot easier. This is where a good sports coach comes in. He keeps his team focused on the game. Different players play different positions, and a well-coached team only thinks of playing the game. They stay very focused! Yes, they make mistakes, but it's how they recover from making the mistakes that matters. Fix the *mistakes*, do not attack the personality, because this will fire up emotions. Bad reactions become very hard to fix. The best way to get respect is to simply and truly listen and understand what people are saying. It's that first step to becoming a good team member, and eventually to be a true leader.

Some people have a hard time because they do not respect themselves, so they in turn don't respect anyone else. This is when it gets hard to deal with. To communicate with these people, you have to try and get them back on track somehow. It's very important then to listen to what they are saying, so you then can remind them of what was said. Actual listening and getting to know people better and accepting them for who they really are is vital because you fundamentally cannot change anyone else. In fact, the only person that can change you is you. So, in a situation or a position of leadership, you will often need to refocus someone. Ask them what their tasks are and what's expected of them as an individual on a team. In line work, a good crew means good teamwork. It's a very dangerous job, so everyone has to watch each other's backs because your first mistake could be your last—and mine was *my* last, for sure, as you will see. I didn't die, but I'm also not doing the work that I loved so much anymore.

I lost my passion and I don't want you to lose your passion. It's not always easy to get back up after you've fallen down. This is why this is such a big deal, to be a part of the team. And how that's done is the coach (or foreman) enforces team play. On a good team, players will help the coach! It's maybe a different way of thinking about it, but it's the best way to think when working with other people. It's all about empowerment. Winning makes you feel good. Doing a good job makes you feel good. Most of all, what we should feel the best about is when no one gets injured and we all go home at night. That has to be the most important thing of all. What's most important to you, wealth or health? You have to make that decision. At the end of the day you want to go home to your loved ones, and if we do that every day, we're going to enjoy our retirement, enjoy our grandkids, and enjoy our great grandkids if we live long enough. It's like they say, live your life like every day is your last, and someday your wish will come true.

[10] MY FIRST CREW

EVEN AFTER ALL I'D LEARNED, I almost had another blow-up, another physical confrontation on a job. This time, I walked off and told myself, *That's enough! I am not doing this anymore. If this is what I have to put up with, I don't want or need this. I am out of here.* I was finished. I had to stop somehow. I didn't want to live like that anymore. I didn't want to get into those confrontations anymore. But old personality defects stay with you for a long time. The way I used to handle problems was causing more problems. That was when I got help from a psychiatrist. I went for about a month and a half, and I learned a lot, but it still wasn't easy to break old habits. Someone would push at me and I would push right back. *Here I go again*, I would think. There had to be a better way. I had come to a turning point in my life, and I had to admit I had a problem.

My psychoanalysis has taught me a lot of life skills, and it wasn't always easy going through that, either. My psychoanalyst would push me to the limit and get me very upset, but I decided I wouldn't do anything and I began to walk away from the session one day: "Paul, stop," he said. I was so angry I felt like kicking the shit out of him. "Paul, you let people push you to the limit and then you blow up like a bomb," he said. "You implode. But what you need to do is tell people when enough is enough, and that they have to stop treating you like that. Tell them you do not want them to talk like that to you, give people warning."

And I learned another way to the control a situation is to just walk away. In the old days, there were no laws about harassment and things often got ugly. Nowadays there are laws for all types of behavior. And I have learned that when people make negative comments, I can simply ask, "Okay, what's going on?" and I tell them I will hear them out. Usually it's just a reaction they're displaying, or a negative comment for control, and if you ask them to comment on what they just said, they don't then know what to say. You call them on their bluff. Sometimes they're even surprised by what they said.

"Principles before personalities" means we do what's right. We don't do things because we are buddies, because the buddy system helps "buddies first," and this is not always the best or right action to take or decision to make. This rule has kept me out of more trouble for the longest long time. Real buddies do what's right anyway, and if your buddy sees you do something wrong and doesn't say anything, you're just setting each other up to fail. Buddies are honest with each other, and if you are really buddies, it can be very easy to succeed. True friendship is very powerful.

I have rules for my work:

1. I will do my job.
2. I don't have to be friends to work with someone.
3. I will not lie for anyone.
4. My good name is on the line. I do not want to fail. And if I cover other's mistakes I am not part of anything good.

I find that at work most people do not want to offend anyone, so if they see someone doing something that is not right, they usually do not want to get on the wrong side of anyone and often say nothing. This is not a good situation to get into, especially on dangerous jobs, but I have seen a lot of it. After a while it just becomes a way of doing business. But what happens is we set ourselves up to fail, because all we are doing is getting into *personalities*. We start just thinking of ourselves and the team dissolves. Everyone simply looks after themselves. You have to re-focus everyone back onto their tasks alone, and not on personalities at all. The team leader has to be strong and this is sometimes hard to do, but I've learned in the long run you stay out of trouble this way. Your reputation remains intact, yet this is the only totally unselfish approach. I have learned to deal with the *problems*, I do not fix the person because I *cannot* fix them. They have to fix that themselves.

I had learned a lot, and it helped me a lot. A year after I quit drinking (and stayed away from it) I got *my own first crew*. I got a job on Vancouver Island, running about 25 miles of line up a mountain. I got clearance from the union and showed up on the first day to meet the supervisor. He was an older lineman named Sam, and he seemed like a nice guy.

"Are you Paul?" he asked.

"Yes."

"Good. I want you to take this setting crew," he said.

"But," I said, "I don't understand. There are a lot of older linemen than me here—"

"No, Paul," he said, "I want *you* to be foreman on this setting crew. You'll do fine. If you have any question I will help you and I'm always around if you need me."

And just like that, I was in charge of my own crew.

Sam was just a really nice supervisor and I didn't need to ask him any questions at all. He checked with me every day just to see how things were coming along. Sam gave me my self-confidence back, and that was exactly what I needed. He helped me by empowering me, and it always feels so good to know someone trusts you. It helped me more than you might think—in fact it's kind of like writing this book, like taking an inventory of myself, it really makes me feel good.

On that job, I helped with organizing and mobilizing tools and equipment, making sure we had all the tools Sam asked for. He in turn made sure we had all the necessary equipment to do our jobs, which after all, was his job as the supervisor. He didn't tell us how to do our jobs, he just enabled us to do our work, and if there was something wrong he would help us fix it. Sam was a real team player. I didn't realize it at the time, but that's what I was a part of—a real team, a good team. We built that line in no time at all and it was all done completely and correctly. Wally, the owner of Surrey Power, gave me a beautiful letter of recommendation that I still have in my possession. It meant more to me than anything at the time, and it was a huge pat on the back. That experience showed me what *good* management can be like, how as a manager you get a better response from your guys, while making people feel good about themselves at the same time. In fact, on that job, I heard no one make any negative comments that I can recall.

One of the best organizers I have worked for was Chuck McCord, on my next job with Catre Hi-Line. Chuck was *not* a micro-manager, not at all, which was very good. I've found that those who end up micro-managing are not good organizers themselves. In fact, Chuck could mobilize and demobilize a crew so well it was a thing to watch. We built the roads with moving equipment, and had no damage to equipment, no cranes or trucks getting stuck, and all while working in the Rocky Mountains. What Chuck did, I believe, that worked so well was carefully pick his crew, and those guys were the best.

Trucks were cleaned and fueled each night, and in the mornings in camp all the lunches were made for the day. All you had to do was get in the trucks, get to work, and get your work done. All very well organized, Chuck was the best I had ever seen. He was all about focus and making things simple. *Keep is simple, stupid!* as they say. It pays because it enables your people, and your employees are the most important asset a company has. Tools and equipment can be replaced, but you cannot replace the practical experience or spirit of your people. I call it "walking experience," while some call it "common sense," or "practical experience." But this means a young worker does not have "common sense" since he has no experience, so we should be careful about our terms. Why curse a young person by suggesting before they have a chance that they have no sense? Words matter, and we want to enable people, not disable them before they even have a chance.

And I learned to manage, little by little, from good examples as well as from bad. I learned we need to simplify things for each other. I used to organize basic things for my crew. At start-up meetings, all I would ask at first is that a crew show up for their jobs on time, and to come prepared for the whole day with clothing, gear, food, and drinks all ready for the day. And after work, make sure your equipment is clean and fueled, check all oil levels, do your housekeeping at night, so to speak. If you find a problem at the end of your shift, it can be fixed before the coming day, which saves us all a lot of headache. I myself would make sure all tools and equipment were ready for next day. Keep it simple. Otherwise one little problem can snowball into a bad day, a bad day into a bad week, and so on. Some people say, "Don't sweat the small stuff," but if you look after the small stuff, the big stuff looks after itself.

I never had to tell too many tradespeople what to do—they know what to do, and they do not like being told what to do, it's an insult. You can stick to keeping in the basics and things can run like clockwork. In fact, if you do have a good, experienced, tradesperson, get them to help you mentor other employees! All the new guys generally need is a little guidance, so help them get up to speed. Mentors are a great way to do this in a non-toxic work environment, anyway. If, on a job, some tradesperson would say to me about another, "He's slow!"

I would say, "Right! Help me with him!"

They'd usually think about it and say, "I don't want to!"

"But you're a very good tradesperson," I'd say, "and I need your help because I can't look after everyone."

And you know what? This approach worked out very well for me. And it made them feel all good about themselves, too. They were helping each other, so they worked together very well. With everyone sticking to their jobs and being productive as a team and starting to look out for each other, we actually all stayed very focused and got along just fine.

I would make sure they understood their job tasks. I would make them do the harder job tasks first thing in the morning because they are not tired yet, they're still fresh and they are more likely to be successful more quickly. You get your harder work done first thing in the morning and at the end of the day organize your crews for the next day. And always know that it's the small stuff that kills the production. Plan on the weather as much as you can. If it's going to be really cold, haul materiel to jobs that are easier in cold weather. Set poles. Manage all aspects of the job and keep crews working. In line work, weather is a big factor. If it's windy and cold you don't want them stringing wire if they don't have to. Do ground work first. But you have to be on the job to do this, not sitting in the office. And you need practical, walking experience.

I ended up working for the City of Calgary for four years and it was a great place to work. I really enjoyed working with the different departments there, and I wanted to get my "cable splice ticket" and I did. I worked downtown on the network system, working secondary wiring, 3000 KVA transformers, and they were all are loop-feeds. The system feeds all of downtown Calgary and is the second largest network system in Canada. Then I worked on a cable splicing crew, splicing 1850 MCM cable (underground oil field cable). It was great work. It was a great learning experience.

I also did commercial work. I wanted to learn the underground part of line work and got my cable splice ticket. This was all a very good experience and I learned a lot doing "hot" work. I have always liked power line work, which is both very challenging, but very dangerous.

PAUL HEBERT

PART THREE: COLD

[11] GOLDEN YEARS

I WAS GOING OUT at the time with nice young lady in Calgary. Her name was Della and she showed me how to ski. At first, we did cross country skiing[15], and I soon found an old style of downhill skiing they called Telemarking[16],

---

[15] Cross country skiing is "a form of skiing where skiers rely on their own locomotion to move across snow-covered terrain, rather than using ski lifts or other forms of assistance" —Wikipedia

[16] "Telemark skiing is . . . named after the Telemark region of Norway, where the discipline originated . . . Telemark skiing gained popularity during the 1970s and 1980s. The appeal of Telemark skiing lies in access - long pieces of synthetic fabric, known as skins, can be attached to the bottom of the skis to allow travel uphill. In addition to allowing a skier to travel both downhill and uphill, Telemark equipment generally weighs less than its alpine counterparts." —Wikipedia

and I did a *lot* of skiing! I skied in Banff, in Lake Louise, and in Fortress Mountain. I took Telemarking skiing lessons and I loved it. I even took an avalanche awareness course, a survival course, and an emergency wilderness first aid class, which is equivalent to EMT training. We did back country skiing and we spent Christmas at Mount Assiniboine. We flew in on a helicopter and stayed over Christmas in a log cabin that had a wood stove for heat. The first night there, the temperature went down to 40 degrees below zero, so it was *extremely* cold in the cabin because they would only heat it up at night, before we went to bed. The hotel guys would come in during the early morning hours and build a fire. I offered to do this for them, as my dad had taught me how to build and manage a fire when I was young.

"I can do this for you," I told the hotel operator.

"Yes, but we are in the national park, and we only have so much wood to burn," he said.

I explained to him how I was going to do it. You build a fire knowing you never want it to go out. You want it to burn very slowly, and I knew how to achieve this. You get the furnace to burn as slow as possible and you close the intake vents as much as you can, so it just gets enough air to burn very slowly. That's the main trick and it saves a lot of wood. You then have a furnace full of red-hot charcoals. This was yet

109

another practical skill I have my parents to thank for. Practical skills are very important, even today. They can be your last means of survival or defense. Yet these skills seem to be fading away today, more and more.

Later on, I finally tried downhill skiing and I fell in love with it right away. I skied every weekend. We used to work nine hours each day with the City of Calgary and we would get every second Friday off, so I went skiing every weekend—some long weekends—and I loved it. For me, it was finding another love in my life, just like when I became a lineman. When you love doing something, it becomes easier to do and you become very good at what you're doing. You eventually become a master at your trade. And while I didn't realize how important this was at the time, it makes you feel good about yourself, too. Line work and skiing *did* make me feel very good about myself. Self-empowerment is so, vitally important. To find *your own* empowerment is key.

But if you're a people-pleaser, some people will not empower you because they cannot even empower themselves. You have to have your own mental power to provide your own empowerment, and in hard times you have to be able to take a self-inventory to see where you need to correct or change. I had to learn how to do my own self-maintenance, to admit what I had done wrong, fix it, and carry on—and to not try to fix other people because they have to do that for themselves. For a long time in my life, I couldn't seem to understand that I could do all of these things and feel good about myself. I couldn't understand how all the time I had spent drinking and partying, while it was fun for a while, was really a waste of time. But when we are engaged in sports, we are usually interacting with other people. And that's wonderful. The attitudes are usually great, and you're all producing the positive, happy juice. I did finally realize that it was up to me, that I have to find positive things, fun things like sports, where I am interacting with people, laughing and having fun. You have to let your own brain heal, at some point, because it feels good. Being happy can be hard to do. I had to work on being good to myself. Don't treat yourself like crap, and don't let others do this, either. You have to find or make your own happiness, but that's okay, because you *can*.

And those were golden years! I was on top of the world and I loved every minute. We did some skiing in the afternoon once it was warmer, and that was actually how I got hooked on skiing. Being on top of mountains like Mount Assiniboine was a once-in-a-lifetime trip, just absolutely beautiful. When you're young it doesn't occur to you that you will probably never go back to someplace like that, but it's in the past now, and I will likely never go there ever again, but I did it. I skied like crazy and I loved every minute of it. I went crazy over it. I was replacing drinking with skiing. And this was where I started being at peace with myself, as I started talking to people, started interacting with people. I just loved it.

I eventually quit working for the City of Calgary (North Canada Power Commission) and went back to work in the Arctic. As "lead line" I really enjoyed working back building lines (overhead lines instead of underground work) and worked in a lot of maintenance, setting poles with hardly any equipment at all. We just used whatever equipment was available. We had a limited number of tools. We set poles using pole-setting pike poles[17], and we would dig a lot of holes by hand because you would dig down just two feet and find it was all permafrost. And this was very hard work. We had to think on our feet and make decisions right there onsite. It was a lot of practical experience that we needed at that time. We would load rock on the truck by hand.

We were a three-man crew and we all worked hard—twelve hours a day for a month and then one week off. By the time your month was up, you were so tired it was hard to work

---

[17] **Pike poles** are long metal-topped wood, aluminum, or fiberglass **poles** used for reaching, holding, or pulling. They are variously used in construction, logging, rescue and recovery, power line maintenance, and firefighting. —Wikipedia

anymore. There's a reality show today that depicts what it was like back then very well, actually, the one that's filmed in Alaska, but we had even less equipment.

This picture is of Eskimo Point (the name has since changed to Arviat Nunavut). Kids loved taking pictures with us, they were very friendly people there. And this was taken in June. If you look to the right, you can see there is still snow!

I was 34 years old in this picture and in good physical condition. I loved my job, and these kids up there were so nice. Notice all the woolen hats they wear! And those big jackets, well, those are summer jackets! I believe the temperature was around 40 degrees *in June* of 1984. Cold! The Hudson Bay is known for its high winds. They call them the "Keewatin," meaning north winds. As you can see, there's nothing on the ground because the wind blows it all away.

By the end of May 1985 this next image shows how much snow was on the ground in Rankin Inlet Nunavut, where we were digging poles out of the snow bank.

In the background to the left you can see the power house which housed diesel-powered generators. It held four power units and they ran 24 hours a day, seven days a week. They hauled fuel by barge in the summer months, and I cannot remember how big the fuel tanks were, but they were big! Big enough that the fuel supplied would last for one full year. They burned around 5,000 gallons of fuel *per day,* which is a *lot* of fuel.

We worked out of Rankin Inlet, which was an Inuit[18] town which had been an old mining town at one time. I believe the town used to close down in the winters because the Hudson Bay is frozen over. The ships can't haul the copper out in the summer months, either, so they only have three months a year for this. I guess in the 1940s they had moved the Inuit people south to Churchill Manitoba. But because the caribou population had diminished, and the Inuit people were starving, they moved the Inuit south to the communities of Wale Cove and Chesterfield Inlet. A lot of people were living off the land, just like they had done for so many years up to that point.

The Inuit had survival skills which young people don't seem to have any use for anymore. But working up north, one wonders how they survived at all in such cold weather. It made me think that it must be survivable (since they did!), since there were trees and grass growing (at least in the summer), but there

---

[18] "The Inuit are a group of culturally similar indigenous peoples inhabiting the Arctic regions of Canada (Northwest Territories, Nunatsiavut, Nunavik, Nunavut, Nunatukavut), Denmark (Greenland), Russia (Siberia) and the United States (Alaska). Inuit means "the people" in the Inuktitut language." —www.crystalinks.com/inuit.html

were actually no trees, only grass. What they did, then, was they ate their food basically raw. They would age their meat to where they call it "sweet." I eventually found out it meant when you can start smelling the meat, which most of us call "rotten." To them, though, it's "sweet." The Inuit do get a good deal of vitamin content out of the meat that way, though. They also boil fish in a big pot of water, or they dry the fish by just hanging it. They don't smoke it, they don't do anything to it, they just eat the dried fish, and this is how they get their vitamin content because the only vegetable they eat is seaweed in the summer. The people who had gotten moved there didn't like living in Churchill, so they moved back because alcohol had taken over, and they did not like it at all. They just simply wanted to live on their own land. They moved back to their settlements and began living like they used to. This was when most settlements were alcohol-free, "dry settlements."

About this time, they installed electrical power and other services into these settlements. I worked up there for four summers and part of one winter, and let me tell you, if you're south of the 49th parallel, you have no idea what's it's all about. In fact, I even worked north of the 60th parallel, which is a long way north, and very cold on the Hudson Bay. It's humid weather and very windy, with sometimes around 100-plus kilometer winds, year 'round. Again, the word "Keewatin" means "north wind" and it lives up to its name. Very windy! I have seen fuel barrels blow around in the street because they used them for garbage barrels. I've seen winds blow up to 130 kilometers an hour (or 90 miles per hour)—hurricane-force winds that would rip the shingles right off the roofs, winds so strong you couldn't go to work. But in those days in the 1980s, it was great money working there. At the end of each 30 days when you had your seven days off, they would fly you home, and it was a six-hour flight home.

[12] THE ARCTIC

WHILE WORKING IN THAT PART of Canada, we traveled to many different settlements like Eskimo Point, Chesterfield Inlet, Whale Cove, Repulse Bay, Coral Harbor, and Baker Lake. We would do line maintenance and new construction work as services were brought to these areas. And it was a great experience, working up north. You often had to improvise if you needed something, as some types of equipment you just didn't have there. And we used all the equipment we could find on hand. I remember there wasn't one settlement that had a backhoe, no picker truck, there was just almost no equipment, at least not like we were used to. All they had were front-end loaders so that's what we set poles with. We would install a steel pipe in the bucket of the truck and hold it in place with guy wire. It was all a matter of thinking on your feet and you had to work smart. Someone called it "working outside of the box," well, there was no box! We made our own box! I learned how to design power lines, how to survey the line, and how to build the line. And of course, we did everything by hand.

I had good practical skills, though, and I was getting better, staying focused on the job. Working up there, you've got the job to do, and that's about it. And this was hard work. It was harsh weather, even in the summer. Some younger linemen had a hard time because they didn't realize that linemen work in the worst weather conditions you can imagine. So, they would make everything personal. I had to lay the ground rules and make them understand that all I wanted was for them to do their jobs. We were all in it together, there with the scant Inuit, and a six-hour flight home. And I did see a number of younger linemen who just couldn't make it because they got too wrapped up and into themselves. There were times I wished their parents had taught them some work etiquette, taught them that work has got nothing to do with your personality, and if you lose your focus, your hard job gets even harder to do! It wasn't even funny, but it's just the way it was. I of course knew I had to stay totally focused.

I had brought along a line buddy of mine. Now, people don't realize how much harder it is to work in arctic weather. The warmest it gets is maybe 40 to 45 degrees during the summer months, and summer is July and August, just two months when there is no snow. And as I mentioned, the wind you have to work in is unbelievable at times. With the wind chill factor, it gets really hard to work. In fact, we usually didn't work in high winds like those at all, but we used work in 50 to 60 mile-per-hour winds all the time. We used to say there are only two seasons in the Arctic—July and August, and winter the rest of the year. It takes a lot of practical experience. If you didn't have it, you go it. And you were always making decisions, hard ones sometimes.

Working in the Arctic, we only had so many months to get all of our work done. The small settlements there had a streetlight on every pole because it's polar bear country. In the winter months, it's dark almost 24 hours a day, so your street lighting has to work all around town because polar bears wandered around those towns in the winter months and it's very dangerous. A polar bear will stalk you and attack you, it's a very dangerous animal, and you simply would not survive a polar bear attack.

The weather conditions were so bad in the Canadian Arctic, and living conditions so rough, you didn't get a room to yourself, you shared rooms. Well, sleeping in the same room as a co-worker (or co-workers) after you've worked all day is not good at all! It adds a lot of stress. The long hours we worked and the long stretches of work (30 days at a time, no days off) were very stressful. We would end up so burned out it wasn't funny by any means. I have seen many burned-out linemen because of not getting enough sleep. In these little indigenous settlements, they often had little camps for us to stay in. We did our own cooking. I would make breakfast because the only meal I knew how to cook was bacon and eggs. We took turns cooking meals and I got things ready to go in the mornings: Up at 5:30 a.m. and at work by 6:30 a.m., off at 7:30 to 8:00 p.m. Long hours, and I could see how hard the physical and mental stress was on the men. Unbelievable, actually, and we had very high occurrence of burn-out.

But we did really well the first year. We set 400 poles by hand. The NCPC (North Canada Power Commission) said they had never seen this much work done, ever. And those were probably the hardest conditions I had ever worked in. We had to stay focused on work, on each task at hand, and on staying out of the personalities, which was the hardest part of it all because guys would get so burned out from lack of rest and from drinking too much that it made it all harder to deal with. It made it hard on everyone and they would just burn out all the faster. I used to tell them that rest was so important because we worked twelve hours per day in windy, cold conditions. Of course, it feels good to release tension after work, but not getting enough sleep wasn't healthy at all.

That's when things got out of hand. I had an old buddy I had hired, a good friend I used to work for on Vancouver Island, and you could see there was something a little off with him right away. Well, bringing him up north like that wasn't a good idea. I knew he was self-medicating, but I didn't realize how much. In fact, toward the end of our time in the Arctic he couldn't function at all. And the guys started going for drinks earlier, at noon. I shut them down, told them to stop the early drinking. I had to, and that wasn't a good situation. We had to let my friend go and he wasn't very happy about this at all. But he had turned into mess, a big ball of emotions, and how do you deal with a person that is out of control? Not so nice things started to happen to him. His wife left him, and he ended up committing suicide.

You run into a lot of people who have problems in this world. Problems were a big part of my life for a long time, but I've never ignored my problems. I had to willfully separate myself from the problems and become part of the solutions, but so many people like to keep doing what they're doing, and it always ends up not good and they pay the price, as my friend did. I had mentioned to him, "Maybe you need to get off the self-meds," and he got very angry. He was very defiant. But what I've seen from my past experience is that too much self-medication makes you very defiant in a very negative way and makes you very reactive. It's a depressant, so how can you be happy taking depressants? You may think it calms you down, but it does not. Some people drink two or three beers and they're happy, but if they drink six or seven they become slobs.

We were also staying in very close quarters and you can't then get away from one another. You have to work together, eat together, sleep in the same bunkhouse together, and it was very hard. Looking back, I wish I could've gone away, but I had no choice at the time, or I believed I had no choice, and I stayed. I stayed for four long seasons, working as foreman, and this was one of the best learning experiences I ever had. I really learned how to improvise, because if something goes wrong, or you need to repair equipment or tools, you figure it out or the job simply stops. Up north you just don't run to the store or go to the line room and get parts or material. You work with what you have. We flew into Repulse Bay to do the line work for the year and we had to build a three-phase line. For a job like that the material is ordered one year in advance. When we got there, we hauled the poles and looked through all the material, and we could not find any rock anchors, and without those anchors you can't build the lines at all.

I had seen an old line that had been built during the DEW line[19] days, which was how they used to patrol the Arctic during the second World War, snipers watching for attack. And I had seen an old line in the tundra which had fallen down. I went back and had a look and sure enough all the rock anchors were there, so we pulled them all up and built new wedges. And we were in business. We didn't throw any material away, it was always reused. And we traveled all over to different settlements, working, as I said. Our line truck was a Beechcraft 18 with twin-turbo prop engines put in it. One time the pilot asked, "How much do your men and tools weigh?"

"Five thousand pounds," I said.

"No way!"

"I weighed everything," I said. "Do you know why we bring these big linemen here?"

---

[19] The Distant Early Warning **Line**, also known as the **DEW Line** or Early Warning **Line**, was a system of radar stations in the far northern Arctic region of Canada, with additional stations along the North Coast and Aleutian Islands of Alaska, in addition to the Faroe Islands, Greenland, and Iceland. —Wikipedia

"No," he said.

"Do you know how much it would cost to fly mules up here?" I asked. He looked at me and he laughed. We were all in very good physical condition, and all of us each over 200 pounds. I was 215 pounds, Derek was around 230, and the other guy was around 220 pounds—all in good physical shape, and we climbed everything with no bucket trucks.

One winter I went to work during the month of January and part of February. It was so cold—in the 40s, sometimes colder—and it was always very windy and very humid. It was what they called "white-out conditions" where *no one* goes out. There was so much snow blowing around you couldn't see and would likely get lost in the blizzard, so they have strobe lights running on a battery system back-up mounted on top of the power house in case the diesel units quit, and the town is without power. Rankin Inlet had a population of about 1,500 people and they had three diesel power units. Electricity in cold weather was everything. Without power, we are lost.

In this photo, the snow blows and goes even higher than the building.

I couldn't believe how much snow there was. You couldn't read the electric meters because they were covered with snow, so you just estimated them. And there were snowdrifts so high that you could just walk right up onto the roofs. The snow was hard like ice, and it stayed like that until the end of June. Loel Olsen was our district manager and supervisor, and I learned a lot from this man. He brought me up through line work, actually. There was a bunch of planning and designing to do and we did our own surveying and designs. I went up there on May 6, 1986, and I had to plan all the work for the upcoming year. And there was a lot to do. We had to build and design a line for Whale Cove, 21 miles of 25KV three-phase to the airport. We had done the survey the year before all the material was there and we were all ready.

126

That spring we surveyed the line and they got a crew together to build the line. They had a four-man crew, worked all summer, and had the line built by October. I had to go and help them finish string the wire. It was a hard job, 21 miles of line to feed the airport electricity so they would have landing lights at night for emergency landings for medical flights. What was amazing was an ice storm came in that spring and totally demolished this line—there were maybe 20 poles left standing. The line was seven months old when the ice storm came through and we measured some of the wire the next day. It was around six inches in diameter, and when it had built up with ice it broke the poles, which looked like they had exploded—unbelievable! I wish I would've taken more pictures. This is when they had to bring in the line crew and had to rebuild the whole line from scratch. Loel and I went to the airport to do a damage assessment, but we just knew the whole line had to be rebuilt, even before we got there.

An old bombardier, with a motor in the back and skis in the front. We had to travel all the way to the airport in this because the line was completely destroyed.

Loel showed me how to plan, design, and survey, which was great for me because in the Northwest Territories (which is called Nunavut) you had to do all your own planning and designs, you had to figure out what size transformers you needed for the load you had, in fact you had to figure out everything. There was no engineer on the job. But that's how we did it in the "old days," you had to do everything yourself.

Transformer pole ice buildup. Picture taken one day after ice storm.

We travelled to Wall Cove and we patrolled the line with an old bombardier, but the damage was so great, there was nothing we could do. Twenty-one miles of line were completely destroyed, and the line wasn't even a year old. About 90 percent of the poles were broken, also. I didn't know how much tension it would take to break a 40-foot pole, but it had to be in the thousands of pounds. And *all* the poles had broken for 20 miles. They looked like matchsticks, cross arms all broken, broken off of the rock mounts. Seeing something like this makes you realize how powerful Mother Nature really is. Nothing can get in its way, it's more powerful than anything on this earth. I have seen this in ice, wind, and lightning, storms. It's unbelievable how powerful Mother Nature is.

We did an assessment. We could not repair the line right away, so they brought in a generator for the airport, and they brought in a line crew to rebuild the line. It took them all summer to rebuild the whole line, everything starting from scratch again. We salvaged whatever we could. I was the lead lineman in the area. We did a lot of maintenance, new services, new poles and transformer banks, and a lot of maintaining streetlights. There was a lot of maintenance to do and it was a lot of work.

The photo that follows is of the midnight sun. This picture was taken flying back from Repulse Bay at two o'clock in the morning, in the month of May. This is on the Arctic Circle. When you're north of the 60th parallel, the midnight sun is north of you. Seeing this, being there, was all very cool.

How often does a person get to see and experience these things?

[13] CARIBOU AND BIRDS

ORKING IN THE ARCTIC you sure have a different feeling. And flying from Repulse Bay to Ranking Inlet is around 720 miles one way— a *long* trip. But we went there on a trouble call one evening. We had to fly to the settlement to turn the power back on. Someone had shot the insulators off a pole. When we arrived and got to work, the job took me about two hours, from about one o'clock in the morning to about three o'clock. Again, long trip! But you're flying in the middle of the night, in the middle of nowhere. You think, *If this plane crashes there's no one there to rescue us. We may or may not survive on our own until someone found us, or a polar bear might find us first.*

Planes up north all fly by beacon because you're too close the North Pole and compasses don't work. If a plane loses signal they know right away that you probably crashed. Very different, living up there. And if you think the summers are hard, well, I stayed up there for six weeks in the winter months one time because Loel went on a holiday, making me the district lineman. Believe me, the conditions up there are so harsh it's unbelievable. You get winds up to 70 or 80 miles per hour, you get blowing snow with the high winds in the Hudson Bay and you can have snowdrifts 20 to 30 feet high. It's unbelievable, and the wind chill is so cold you don't even turn the power back on because you can't get there to fix it. If the power goes out in a house, they have a drain system. They drain all the water from the home and when the storm subsides you turn the power back on because it's strong winds and maybe 40 to 50 degrees below zero—with the wind chill it's probably closer to the 60s and 70s below zero. You cannot survive in weather like this. You stay inside. You cannot climb poles in high winds like this.

But you see the locals and they're outside in 40 to 45 degrees below zero with a light wind, and they're used to it. Unbelievable. The dog stays outside year-round! They dig themselves into the snow and that's how they survive. We think it's really cold in the southern provinces, but I traveled to Winnipeg at one time and it was *negative* 52 degrees in Rankin. When I got off the plane in Winnipeg it was 38 below and it actually felt warm.

Loel Olsen and I, without any planners or engineers, started working in the Arctic and we had to design and build things ourselves. In the old days, we just built power lines, sometimes without any surveys or stakes. They started building a lot of houses and there was no surveying done so they had the property lines and the streets all in the wrong places. It was all messed up, so we had to resurvey everything, stake everything, redesign all the power lines, and build the power lines in the right locations. It was interesting though, and I learned a lot from Loel.

This is a picture of Upper Falls on South Hampton Island at the northern tip of Hudson Bay. Absolutely beautiful.

Someone suggested if we would like to see Upper Falls we could make the trip on quads (four-wheelers). We traveled for about three hours across the tundra, leaving around 4:30 a.m., and we didn't get there until about nine o'clock or so. It was a *long* ride. But once we got there it was absolutely beautiful, amazing, in fact. And we saw all kinds of things. There was an old DC3 airplane that had crashed in the 1950s. It looked as if it was brand-new. The Arctic minerals had preserved it so. And with little heat from the sun, the paint hadn't faded apparently at all. It was like the day it crashed, it was unbelievable. Then we carried on to Upper Falls and it was just absolutely amazing. The sun was shining so bright and the sky so clear. No pollution, just beautiful, clear, and quiet. We saw caribou and birds. This was at night around 11 o'clock, and you could see for miles. It's the way it has always has been and it's never changed up there—except maybe for the plane! But that's the way Mother Nature made it, and it was still the same when we were there. Beautiful.

We would always finish at the end of September because the weather was just too harsh and that season we had all our work done, as usual. I went home, and I worked in the ski area called Fortress Mountain and I just loved it. The two things I loved most in life were building power lines and teaching skiing.

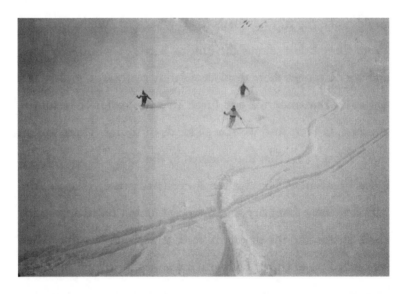

One of my loves in life was skiing, especially heli-skiing. I am the skier

on the left

I think this is what heaven looks like, I thought. I loved everything about those golden years of working and skiing. And I *loved* to heli-ski (aided by a helicopter rather than a ski lift). If I could, I would do it all over again. You're standing on top of a mountain probably 8,000 feet high, and you're looking down. Below it's just absolutely beautiful. I have to be grateful today for the things I did, things I am not going to do ever again. We're only going through life once and whoever has the best time is the winner, I guess. I chose my destiny; we choose our actions, choose our words, we try to choose our friends, and we try to be nice. Sometimes it doesn't always work, but we do the best we can. I got to be pretty good at skiing. And I was good at teaching skiing.

I had a great opportunity with it, too. Kobe Weise with CMH Heli Skiing wanted me to go work as a winter guide for heli-skiing groups. I had skied with Kobe, so he knew I could do the job. I had my Level 1 avalanche awareness and I had taken my emergency wilderness first aid as well (although I didn't pass the emergency wilderness first aid, missing the grade by just two points). And by this time Fortress Mountain was calling me to go to work for them as a ski school supervisor, so I was torn between heli-skiing or working at the mountain. I chose to go work at the mountain. Now that I look back, I wish I would've gone for the winter guide job—that was an awesome and even unbelievable opportunity—but I didn't, even though it was as big a dream as working on power lines. I really enjoyed it. I loved the outdoors. It gave me such a sense of freedom to be out in the wilderness enjoying the beauty in its natural form. There was never a dual moment.

My other dream was to become a Level 3 international ski instructor, and I was so close—all I had left to do was go for my two-week course and take my Level 3 international test. If I would not have had my accident, I would've done it. But I'm very grateful I got to do these things. I did what I wanted, and I was very happy doing it, and today I'm still grateful. I got to see the world from a 9,000-foot mountain peak, and the world looks very different from up there—so big, so beautiful. It's actually very spiritual to see the world the way it is. I like the native saying that there is really nothing wrong with the world, it's just the people on earth that have problems.

Fortress Mountain is absolutely beautiful. In the photo that follows, you can see the ski runs on the backside and to the right is the continental divide. We used to get 10 or 15 feet of snow for great spring skiing. Today Fortress Mountain is closed as a ski resort, I think because the road up to the mountain needed a lot work, so now they only do snow cat skiing.

At Fortress Mountain, I worked with a lot of nice guys. Graham Smith, for example, was a very nice guy and a great skier. And I liked everything about being part of a ski area. I worked for a school director named Don Dairy, who was a Level 4 instructor. I worked with some Australians too, and they were great people, we had a lot of fun. I ended up working as a ski school supervisor and I did my job well. I enjoyed everything about it.

Joe Collard was a great guy and a self-made millionaire. He started the first ski hill in Calgary called Pascapu. When he first opened people wondered how because there was hardly any snow in Calgary. Most had never heard of *making* snow, but that's what he did there, and it was great success. He had one of the biggest ski schools in Canada, actually. I heard that some of his school groups were well over 1,000 students. Joe later sold it to the Calgary Olympics for the 1988 Olympics and he purchased Fortress Mountain.

He wanted me to come and work for him, fixing the ski lifts in the summer months because I was good at working at heights. I had very good practical skills. Looking back, I regret not going to work for him. He told me he was going to bring me to higher places and that he was going to look after me if I would go to work for him. I didn't, but I have had such a good life, I have no regrets. I was good at my job, and it seemed that

I always liked what I did. I always wanted to do things to the best of my ability. My dad was a tough dad, but he taught me a lot, including that you have to respect yourself first, and if you do that it's a lot easier to work with others.

I have always liked to tell jokes and always liked laughing. I had to learn how to survive, and laughter is how I did it. To this day I love laughing. I think of this in terms of dogs: One dog will be wagging his tail, just happy to see you; another dog might be barking, growling, he's not a nice dog. People are very much like this. You have to watch for nasty dogs because they are dangerous. If you meet people that are nice or pleasant, happy to see you, those are good people, but then you meet people that are angry, that are loud and boisterous. These are the ones I had to learn to stay away from. You cannot deal with these people, they're just plain angry, and they have to empower themselves with anger because they're full of fear. Fear is a byproduct of anger.

Very insecure people have a hard time staying focused. They get into the personalities and emotions of situations and they often only talk about themselves. On the other hand, happy people will talk about you, not just themselves. They want to know about others and they don't talk *about* people, they talk about productive events and they tend to like to look into the future. They're visionary. They want to discover good things about people. The negative stuff they don't care about.

I finished working in the Arctic after four seasons in 1987, stationed out of Ranking Inlet. I remember the first year, Don McDonald said that we had done what he had never seen done before, we set 400 poles with very limited equipment. I was pretty proud of what we accomplished there. Loel Olsen had helped me with the planning and designing and together we surveyed and designed the structures. I learned a lot working in the Arctic, and I learned how to improvise. I loved the hands-on experience even though it was very stressful. I had seen so many people burned out, but it helped me remember to stay rested and to not drink. I had to keep my strength and my energy for long hours and it was extremely cold even in the summers, but I maintained myself and I was very successful.

There was one guy who I will always remember and respect. He stuck with me all the way and we are still good friends. We worked well together. His name is Tom Caruthers, a great lineman and a great guy. Another great lineman was Tom McDuff, as well as Derek Chudyk. These three guys really stood out, and I loved working with them. We often ended up doing the work of a four-man crew with the three of us. And it kept us away from the other, toxic personalities, which made it so much easier to do our jobs.

I left the Arctic and went back to Alberta, burned out as I was by then. My supervisor Bill Burk from North Canada Power Commission recommended me to Victor Budzinski, the owner of Valard Construction. Victor called me at home and asked me if I would come to work for him, and I enjoyed talking with him. I went to Grand Prairie and met Victor, and I went to work as a lineman. I didn't really want to be foreman, though, because I had had enough. I worked for about two weeks and I was enjoying every moment just doing my job.

After about two weeks Victor wanted me to go along to another job to a line rebuild. They were opening a school in Jean Cote, a small-town in Northern Alberta. I ended up staying in my hometown, Falher Alberta, where I was born and raised. At first, it made me question what I was going back to work there for! But I actually liked it, because I got to see people I hadn't seen in over 20 years. It was great. And this was actually where I met Lorraine again. I didn't think anything about it at the time, however. I used to eat at Lorraine's restaurant every day because it was the best restaurant in town, and she's a great cook and a great lady. The following winter I left to work in Rycroft, and then in Grande Cash, Hinton, Alberta. And after those towns, I decided it was time for a change for a while.

The clear, beautiful, winter sky at Obed coal mine. We moved 3-phase line for the barge lines.

[14] HAWAII MONTHS AWAY

"I'M GOING BACK TO WORK as a ski instructor," I told Victor Budzinski, the owner of Valard Construction, and he didn't want me to. He had bid on the transmission line in British Columbia, and he wanted me to run the job. Despite getting all the paperwork done and being pretty sure he was going to get the job, it didn't come through. I went back to work on distribution lines in Rainbow Lake. That winter we also ended up working in Fort Vermilion, then back to Rainbow Lake again. We got that job finished and I got work in my hometown, building a three-phase line and doing maintenance work for Alberta Power. And back home, one thing led to another. Lorraine (I was a regular at her restaurant) and I started talking a lot. That led to dating and that led to getting married on September 23, 1989. And we were on top of world. Lorraine had a good business, I had a good job. I was working as a contractor for Valard Construction again, making very good money. In fact, we used to ask ourselves, "How can life be better? We have everything—we have each other, we have good lives, and we have a good income." It was so good it was unbelievable.

Alberta Power wanted to hire me in Peace River, but Victor didn't want me to leave. "Stay," he said. "I will look after you, just stay with me. You can join the company, in fact." And I did. I almost went into the line business on my own, actually, soon thereafter. There was a small line contracting company I was tempted to buy.

146

Lorraine and me. The happiest day of my life, we had everything!

In the utility business, it's always very busy before Christmas because they're trying to finish the year-end contracts, so there was a lot of work. I told Lorraine we should go on a honeymoon after Christmas because the company shuts down for two weeks, and we planned to go to Hawaii, and we were really looking forward to it. But Hawaii was months away. It was September 30, 1989. We had been married just one week. I had a "planned outage," to move a three-phase line for a new building. I got up that morning and I looked outside at about six o'clock a.m., and there was about six inches of wet snow. It was still snowing and raining, it was windy too, and cold—brutally cold. I told myself I would never get the equipment I needed for setting new poles and for salvaging the old line and the transformer bank I had to move, because it was also just too muddy. I called the control center and cancelled the outage. Lorraine had to go to open the restaurant, so I kissed her goodbye and went back to sleep because I was tired. I had been working a lot that summer, and on a lot of trouble calls. The phone rang at about nine o'clock a.m. I answered, and it was the control center.

"The assistant district manager is out restoring power and needs help," the man said. "Can you go out there?"

And in the line business, you just cannot refuse to go to work because when power is off, it's classified as a state of emergency, so it's your duty to go.

"Yes," I said, "and if you have other linemen who can help, please send them." Then I called the ADM (assistant district manager), a man named Blaine with Alberta Power, and asked him where I needed to go to restore power to customers. We proceeded to the first location and the conductor was down. We opened the switches and we did our clearance. We removed the hot clamp. We installed grounds, which was the old procedure, so we thought we were safe. We repaired three breaks then called the ADM to see where more breaks might be, and he told us about the next location.

"Just go there," Blaine said. "I'm gassing up my truck and will meet you there shortly."

Once at the second site, we found a break in the line, and the wire was crossing the road. We installed our grounds, cleared the mainline, and repaired the span of wire crossing the road. The ADM showed up.

"If you remove your grounds and give me your clearance," Blaine said, "I will go re-energize the line for you." I did our procedure, and when we had completed our work I made sure all our men were in the clear.

"We're in the clear and will stay in the clear," I said. "It's okay for you to re-energize the line."

"Okay," said the ADM, "consider it hot!" And he left.

But he soon came back. "The fuse didn't hold," he said. "There must be one more break in the line." I went down the road looking, and around some trees, sure enough the line was down. I reported to the ADM.

"I'll go do the switching and grounding for you," Blaine said.

"That would be great," I said. After a short while, he returned. "Are we all clear to go to work?" I asked.

"Yes," Blaine said. This ADM was with Alberta Power, so I didn't question him. It would be the worst mistake I would ever make in my life. I had *doubt*—the most powerful instinct we have, but I didn't trust it.

Rick, my pole partner, went up the pole, and he dropped a span of wire down to me because it was easier to splice on the ground. I installed a cable jack and two grips on the wire. I started pulling the wire together and that feeling of doubt came back—I just didn't feel right at all.

I turned to Blaine, the ADM. "Did you leave your ground back there?" I asked.

"No!" he said.

"Did you take the hotline clamp off the main line?" I asked.

"No," he said.

"Do you mean to say I am working an open switch? Wow! I hate working like this!"

And at that moment the power came on. The line became energized at 14,400 volts.

PART FOUR: HOT

[15] God, Help Me

WHEN YOU GET HIT by such high voltage you see nothing but white. There's no smoke—the intense heat burns *everything*. I felt the power, the boom, the high-voltage arc, I remember it all. When you make contact with high voltage like that you can actually hear the power generators from the power plant—it's unbelievable. And it all happened in a tiny fraction of a second. The heat from a high voltage arc is 37,000 degrees—comparable to a nuclear blast, and it blew me back more than 20 feet. I remember laying there—I could hear everything, but I wasn't breathing, and I could feel that my heart wasn't beating, either.

This can't be happening, I thought. *I've only been married for week! I'm a dead man.*

I heard Rick, my pole partner (a big man, about six-foot four) screaming at me, "Paul are you alright? Are you alright?" He grabbed me by the front of my jacket and he shook me. I felt my heart start to beat again, I started breathing again. (A doctor later told me had Rick not done this, I would be dead. He defibrillated my heart. Rick saved my life and I am very grateful. It's really hard to understand what Rick must have gone through that day, but the sad part is I have tried to talk to Rick and he doesn't want to talk about it at all. But he saved my life, and again, I am very grateful. *Rick, if you're reading this, thank you very much. I owe you big time. I would love to talk to you someday. You did the right thing, bud.*

My body went into a survival mode, called *post-traumatic stress.* You produce really high levels of adrenaline—it's a huge form of panic. Our body does what it has to do to survive, and you don't really understand it until you have been there. And at the moment, what was happening to me did not seem real. You think it's all a dream—you hope it's all a dream—what's going on, you don't really know. It happens so fast it's unbelievable. You're not thinking of material things—money or even people—*you just want to live.* And I didn't want to die in that ditch.

I don't care if you believe in God or not, I was overcome by a feeling of *powerlessness*, and the first words that came out of my mouth were, "God, help me." People would later say God must have had a lot to do with my incident, but I don't think God had anything to do with it at all. We simply do these things to ourselves. We are not perfect, we make mistakes. Despite all of the guidance we get and our own experience, sometimes we just don't use it or don't pay attention, and this is when we become "invincible." I call it "easing God out." I only remember that I didn't want to die. I just wanted to live. I wanted to see Lorraine again—we had been married for just a week! And we were so happy—that was all I was thinking of. I thought I would never see her again. But what did happen was a miracle. They say God works through people, and in mysterious ways. I say God wouldn't do things like that, that God does not hurt people. I don't believe he does. I like to say God had a lot to do with saving my life, but all I know is lying there, desperate, I uttered, "God, help me."

When I started breathing again the pain was unbearable. I looked at my chest expecting flames but there were none—just indescribable pain. I would later find, from Doctor Tredget, that linemen survive shocks—*when* they survive high voltage— because they are usually in good shape and have low heart rates. But I wasn't sure I'd survive in those moments. My hands were curled closed and as much as I wanted to open them I could not. I was in a bubble of unbearable heat from the blast, and it burned for a long time—to this day I'll sweat practically all of the time because the burns seal your skin and you just don't cool off like you used to.

I looked over and Blaine, another lineman with us, was lying on the other side of the ditch face down in the mud and snow, and there were gurgling sounds coming out of him—not good at all, I thought—like he was already dead.

"Rick," I said, "do you know first aid or CPR?"

"No."

"Roll him over on his back," I said, "and lift his hands above his head as high as you can. Hit him in the chest three times then press on his chest three times. Lift his hands above his head again, and just keep repeating this."

Rick did. The gurgling sounds came more and more from the lineman, and finally he came to. It seemed like it took forever to revive him. As he came to, he was growling and making noise like a wounded bear. I remember him putting mud in his mouth trying to cool the inside of it, putting mud all over his mouth and face. It was incredible to live through, and it's incredible—*extremely difficult*—to write about this now! I remember the panic I was going through in my mind—I thought my life was coming to an end. *We're not going to make it,* I kept thinking, *It's over.* I didn't think I was going to see my family again, and I had been married just a week.

And talk about miracles! I had worked for Hydro One in Ontario and every Friday we had a safety meeting. They didn't believe in CPR, so they made us practice the Heimlich maneuver and believe me, this is what saved Blain, the other lineman. These are such important life skills, and when they're used successfully, we call them *miracles.* I had taken my Wilderness Emergency First Aid training, a very advanced first aid, and I think it helped me understand that we had to get to a hospital as fast as possible.

Rick went over to the truck, picked up the radio, pushed the star key and shouted "Mayday! Mayday!"

The control center answered. "What's the nature of your emergency?"

"I have two linemen *electrocuted,*" Rick said. "I need an ambulance."

"What is your location?"

"I don't know—Paul?" He was asking *me?* "Paul, where are we?"

"By the Marcel Clottier farm," I said. It was approximate, but it was a start. They made sense of this and sent an ambulance out to our location. But a red and white pickup truck was coming our way and I hollered as loud as I could and he stopped. I knew the guy—his name was Lucian. "We've been electrocuted," I said. "We have no broken bones, but we have to get to a hospital as fast as possible."

"What do I do?" he said.

"Put us in the pickup truck and take us to hospital!" I said. "The worst thing that could happen is our hearts could stop. If that happens just hit me in the chest to restart my heart. But we have to get to the hospital as fast as possible."

I have since talked to other linemen that got "electrically socked," which is the proper terminology, and it's not easy for others to understand. If I could've taken all my clothes off in that ditch I would have because I was so hot, and I didn't know what to do. My hands were in the fetal position—I couldn't get them moving at all. I couldn't move my feet. I couldn't do *anything*, and the burning and the pain was so bad all we did was scream. It's unbelievable what you can live through.

Once they put me in the pickup truck to take me to the hospital the ambulance arrived.

"Do you want to go in the ambulance?" they asked.

"No," I said, "just take me to hospital."

It was not the right thing to do but the EMT was okay with this because he couldn't see the damage I had suffered yet. But if my heart would've stopped again the EMT could have restarted it *if I had gotten into the ambulance!* But I was anxious to get to the hospital because I wanted to see my wife again just one more time—that's *all* I wanted. I wanted to live, but I knew I was in a lot of trouble.

The first thing they did at the hospital was cut all my clothes off. They were bringing ice packs out of the ice machine and they were icing my body all over—my legs, chest, arms—because I was so hot they had to cool me down. I was confused by all of it. They put a catheter in me, releasing blood from muscle fiber, from the damage from the electrocution, which is what destroys your kidneys if not handled. They gave me painkillers—I asked for more until the doctor said he couldn't give any more. The doctor was walking around telling the nurses what to do next because they had to get me ready to fly me to the burn unit. I told them to call Lorraine.

And when Lorraine showed up I *so* happy to see her because I truly did not think I would ever see her again. And I thought *that* would the last time I would see her, there in that hospital. I was telling her about bank accounts because we had just gotten married, and we hadn't merged our bank accounts.

It's unbelievable to be so close to death. I'm crying as I write this, but I'm still here.

They drove me by ambulance to the airport and put me in a small medevac[20] plane with a team of nurses and a doctor so they could revive me if they had to. There was a nurse there I had gone to school with named Marilyn. She was talking to me and they were giving me pain medication.

"Paul," she said, "close your eyes, relax."

"I don't want to close my eyes," I said. "I'm afraid I'm going to die."

I told her a lot of things to tell Lorraine, who was and still is the most important person in the world to me. I was so afraid.

When we landed in Edmonton I couldn't hold on anymore and everything went dark. I blacked out. I woke up in the burn unit. I barely remember anything around that time, but when I woke up again two days later, my left foot had been amputated as well as two fingers of my left hand. I was in a daze. I didn't know if this was all real or not—I was confused. They had me on lot of pain medication, but I still had a lot of pain. Lorraine was there, telling me how they had to amputate my left foot

---

[20] "transport (someone) to the hospital in a helicopter or airplane" — Google

and two fingers. The doctors had instructed Lorraine to talk to me about my amputations, so I would accept what I was going through. And I can't really describe it. It's a feeling of losing yourself and losing what you did in your life. And I became scared Lorraine might leave! You worry about what you're going to do, how you're going to survive, how you're going to make a living. I had a foot missing and most of my left-hand and I would get delirious, fearing the burned skin might turn gangrene, turn toxic, and I would start losing my mind. I started hallucinating, having very bad dreams. I woke up in a butcher shop just to find it was a bad hallucination. I would wake up and worry until Lorraine came back the next morning.

And you don't think about this except with counseling. I didn't know at the time, but these were panic attacks and fear. I was so afraid of being alone. And it would all turn into deep depression.

I begged and pleaded with the doctors to save my right hand, but it wasn't going well, and they had to remove all of the damaged flesh before they could amputate my right hand. I got delirious, and they put me back in surgery and removed the damaged skin and muscles. It was a brutal process. They didn't induce coma back then, which nowadays saves a lot of energy and it eliminates a lot of pain—pain I lived through, despite the morphine. I would hallucinate so bad, I didn't want to take the meds anymore, but the doctor insisted.

"Paul," he said, "if you don't take the pain meds there's going to be too much pain and your heart wouldn't take it, you would not survive. You have to take the pain medication."

I did. They switched me to Fentanyl[21]—very powerful, like anesthetic and it works instantly. But I still always had pain, it didn't ever go away completely. Electrocution and bone cancer are the most painful things you can survive, at least that's what Doctor Tredget told me after I had completed all of my surgeries. I was going in and out of surgery and it was unbelievable. They had to stop giving me pain meds just before I went into each surgery, and every time I came out of surgery it was so painful. They had to rebalance and build the pain meds back up in me to control the pain. In the hospital there, all together I had around 40 hours of surgeries.

What I was taking and the side effects of all of this were unbelievable. My heart rate used to go up to 180. When my brother come to see me, he took his hands and he put his fingers around my forearm. He couldn't believe how much weight I had lost after all the surgeries were done. (Today I

---

[21] "Fentanyl (also known as fentanil) is a potent, synthetic opioid pain medication with a rapid onset and short duration of action." —Wikipedia

have had over 70 hours of surgeries.) With all this pain medication I was so constipated, I pushed so hard to go to the bathroom my stomach wall ripped open. I freaked out, of course, and had another bad panic episode. The mental part of it all really scared the hell out of me. I didn't know how to deal with those thoughts, those fears you go through. I was so afraid through it all. I had so much pain I didn't want to go on anymore, I just wanted to die. And even after all of the physical pain, the mental pain is the hardest, I think, that I have ever lived through.

To think, when people are being told and taught to apply safety procedures, you hear people say stuff like, "Safety is a bunch of bullshit." Well, it just drives me crazy to hear that. How ignorant! You can be so uneducated as to think you're invincible, that there are people out there that do not think "it's ever going to happen to them." It's a cavalier, fearless, *dangerous* attitude. Ask yourself what you would do if you were in my position. I was thinking of a lot back there in that hospital. I blamed myself. I hated myself for making a bad mistake, and I am still paying the price, then, now, and for the rest of my life. It's hell on earth.

They debrided[22] my right hand *and* my right leg because they had to take the burnt flesh off to get to the good flesh so they could see how much damage there was. All the time I was begging and pleading with them—I didn't want to lose my right hand! I knew if I could keep my right hand I could do a lot. One day they came in and removed all the bandages and *showed* me my hand. It was all dry skin and dry bones, hand and wrist. The doctor looked at me and I knew what he was going to tell me.

"I'm sorry Paul," he said. "We have to amputate your hand."

It's a feeling hard to describe, hearing that. Empty, losing more and more. Major depression was setting in. I didn't want to live anymore. I was losing a part of myself. Losing my hand was very demoralizing. I had never been in that depth of a depression before and it was so hard, it was so bad, it would have felt better to die. I had lost my identity. I was nothing. I couldn't even wipe my own butt—how humiliating is that? I really felt my life was over.

---

[22] "To remove dead, contaminated, or adherent tissue and/or foreign material. To debride a wound is to remove all materials that may promote infection and impede healing. This may be done by enzymes (as with proteolytic enzymes), mechanical methods (as in a whirlpool), or sharp debridement (using instruments)."
— www.medicinenet.com/script/main/art.asp?articlekey=40481

[16] LORRAINE

I DIDN'T WANT TO GO back into surgery but I had to. They told me there was nothing they could do about it. In the operating room, Doctor Tredget was looking at me and talking to me and I said, "Doctor, why are we doing all these surgeries? In the end aren't I going to die anyways? Why don't you just let me die in peace and stop the suffering? I don't want to suffer anymore. I have had enough. I don't want to go through this anymore. I just can't—please just let me go. I just want it to stop!" I was crying, I just couldn't do it anymore. I didn't *want* to. I didn't want to lose my right hand. I didn't want to lose parts of my body anymore—it's very hard losing parts of yourself, one piece at a time. Even 26 years later, I'm writing this, and it's still very hard to talk about. Losing limbs is the hardest thing in life. Doctor Tredget told everyone to leave the operating room and he sat beside me on the operating table. He reassured me.

"You have a great family, Paul. They really care for you, they want you to hang in there and they do *not* want you to die," he said. "Paul, do it for Lorraine. She loves you very much. You are *not* going to die. You're going to be okay because you're a strong person and you have strong people backing you up. You are going to be alright, you will make it."

He convinced me to go back into surgery, and that's when they amputated my right hand. Every time I came out of surgery Lorraine was always waiting there, and that meant so much to me. She has done a lot for me and I think this actually brought us closer. Today we are so close. At one time, she wanted to go home because she had a restaurant to run and family at home, but the doctor told her not to leave because, he said, it was a good possibility that I probably wouldn't make it, if she would have gone home and left the hospital. Lorraine stayed, and she was there every day. She saved my life. Family support is very important.

But I wasn't the only person that was hurting—your friends are, your family and co-workers are too. In fact, your family goes through an absolutely helplessness situation. When they can't help, they feel so hopeless. All they can give is moral support, that's all they can do. But this means a lot to the person in need. I felt very hopeless, so Lorraine just being with me was everything. She and the other people that visited me saved me. I didn't realize how many friends I had. If you have a friend on their deathbed, please, go see them. At the end money and material things don't mean anything, but your friends and your loved ones do, and believe me, that is what kept me going—Lorraine and my family and my friends.

After they amputated my right hand I was feeling better. You sometimes don't know what the outcome of something might be. All I kept thinking was how I wanted to be able to ski again, because I just loved skiing, just loved it. *Heaven must look like the top of the mountains.* I was supposed to manage the ski school the winter before my accident, but of course all of that changed. My plans were just to try and survive. And through all of this—I didn't know it at the time—I was going into a mode if survival, going into post-traumatic stress.

They thought they could save my right leg if they operated again. The surgery was called a "flap transfer," and Doctor Tredget said they were going to take a muscle off of my back, off of my left shoulder, and rebuild the calf muscle in my right leg. They said it would be an eight-hour surgery that had a success rate of about 50 percent. Fifty percent was huge for me! I thought if I could save my right leg I could probably walk pretty well, and I knew I could ski on one good leg. They did all the preparation before the surgery and I went in very hopeful that the outcome would be good. The surgery lasted 15 hours instead of eight. When I came out I was really worn out and in terrible pain because I was completely off the pain meds. It was brutal. Lorraine and my boss Victor came in after the surgery to see me.

"How are you doing, Paul?" Victor asked, but I was in a terrible state of mind.

I told him, "I just don't want to talk to anyone, Victor. I just want to be left alone." So, they left.

I had given up completely. Paulette, a nurse I had known in school, was working in the burn unit, and I had a Hemovac[23] in my back, draining the fluid from the muscle. It was so uncomfortable, lying on this drain tube, and I had so much pain. I was so uncomfortable, so restless, and in excruciating pain.

---

[23] "A Hemovac drain is placed under your skin during surgery. This drain removes any blood or other fluids that might build up in this area. You can go home with the drain still in place. Your nurse will tell you how often you need to empty the drain."
— https://medlineplus.gov/ency/patientinstructions/000038.htm

"Paul, can I do anything for you?" she asked. She had been giving me pain meds.

"Paulette," I said, "thank you, but if you could bring me a gun I would end this right now. I don't want to live anymore."

Again, Doctor Tredget came in and reassured me. He was such a nice man. He calmed me down and got the nurse to get some meds. "Just take it easy, Paul," he said. "We've sewn an IV drip into your artery to control the pain. We cannot skin graft your leg right away, we have to make sure we have circulation first." They were putting plasma in me, and the plasma was really cold, so I was freezing. I was in so much pain I can't even describe it—again, I'm crying as I write this. And the pain is still there. I don't think it will ever go away—you just learn how to live with it.

The doctors came in and they took the tensor (elastic) bandages off. After looking my leg over, they said we had circulation and they were going to graph the leg. Into surgery I went for another three hours of skin grafting. I came out of there weaker than ever, and even more depressed. I couldn't even lift my head off the pillow. I was down to 140 pounds (I weighed 215 pounds before the incident). I was always a strong guy, very physically fit, and I just couldn't believe what I was going through, and the pain. I was losing everything. They said they couldn't disturb the leg for three days, and three days later the doctor came in and removed the tensors and bandages. Right away I could smell rotten meat. The doctor looked at me and put his hand on my shoulder.

"I'm sorry Paul," he said. "I'm very sorry, but your leg has to be amputated."

I lost it. I didn't want to live at all. I sunk into a depression. I wished that somehow, I could've ended it or died. But they didn't waste any time. They very quickly had me ready for surgery and amputated my right foot. I didn't want that surgery at all, and I cried all the way to the operating room. Lorraine was there, talking to me, but I still didn't want it. I know what it feels like to want to die—*Please let me die, please just let me go in peace! I have had enough!* I will never criticize a person for committing suicide, not after what happened to me. I couldn't take the pain anymore. The doctors made my choice for me that day because they cannot let you die. I would be dead today if I would have had my way. How many people go through something like this? I know what it's like to cross that line—it's a very empty feeling. It's too much pain, it's very high stress levels. I call it an overdose of stress. So today, I just don't stress anymore. I hate stress. I have had enough. Today, if people are angry and stress out I just get away from them. Anger is just a bad defense, and an illusion of empowerment.

They were giving me medication to control the pain and my heart rate, but it wasn't working anymore. My heart rate was going through the roof, around 180 or so. They had to bring it down somehow, so they gave me a shot of valium to calm me down, and it didn't work. I felt very weird. I defecated in the bed. I lost control of everything. All I can remember is the nurses giving me shots of something and hollering at me, "Paul! Stay with us! Paul! Wake up! Stay with us!" They were slapping me, reviving me. I came to and I was apologizing because I had crapped all over the bed.

But what had actually happened was *I had died,* and they revived me. It was such a *peaceful* feeling. As I think about it now, I wish I would've never came back. I will always remember that feeling, it was unbelievable. People ask about dying, about how it feels. It's the most peaceful thing I've ever experienced in my life. My psychologist, years later, asked me, "Paul, are you afraid of dying?"

"No," I said.

He explained, "Paul, people who have had near-death experiences, or people who actually did die and were revived are no longer afraid of death." And I believe it. I have never felt anything that peaceful in my life, ever. It is very true that you are finally at peace—total peace. No one can harm you there. It only hurts those you leave behind.

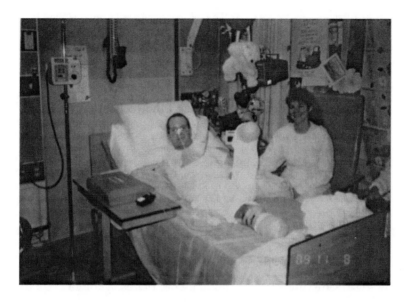

Lorraine comforting me at the hospital.

I don't know if there's a heaven or what there is, but I strongly believe there is *something*. And one thing that I'm very grateful for is my wife Lorraine. She never left my side in the hospital, and I hung onto her. I would wait for her to arrive. She was everything. She's still everything, and when she's not around I really miss her. And you know, how I survived, it changed everything. It changed how I think. I actually didn't know what was going on most of that time, not really, not until I spoke about it with a nurse I know. And I am no longer afraid of death or dying. I died twice, once in a ditch. We are all going at some point, and all I can say is, it was very peaceful. Of course, the whole experience sure made me think, and it changed the way I think and live now. My fear now is I never want to be alone, it just eats me up.

Once the surgeries were all done I did feel better. As I said, I was not even sure what was going on, in fact I had no idea, but Lorraine was there, and she kept reassuring me.

"Don't worry, Paul," she'd say, "it will be okay. We'll be fine, don't worry."

She would put me in my wheelchair and bring me to the dining room. She would bring me chocolates. I couldn't go to the bathroom by myself, I had to get the nurses to help me. And let me tell you, a person like me, who was so independent all my life, who didn't get married until I was 40, suddenly found myself totally dependent, and that was hard. It was humiliating. I felt as useless as a baby. They had to brush my teeth for me because my left hand wouldn't function. They had to shave me. They had to wash me. It was totally embarrassing and humiliating. The fear of the unknown sets in and sets in hard. *What's going to happen to me?* I felt constant panic, constant powerlessness. My boss, Victor Budzinski, came to visit me often. He was a big part of the moral support, a big factor in my recovery. I call Victor "the man with a very big heart." He never abandoned me, in fact he looked after me, that generous man. He's still nice today, by the way! We're still very good friends and I'm very grateful.

They were taking me off the pain meds and I had some pretty big withdrawals from the Fentanyl and the morphine. I would cry a lot and get very emotional. I was very depressed, and more and more reality was setting in. Lorraine wheeled me into the hospital chapel one day. There was an Oriental lady playing the organ and it started me crying and I just couldn't stop, it was just pouring out of me. I think they were tears of healing, though.

"Can I play something for you?" she asked.

I remembered, when one of my best friends had died, his favorite song was Ave Maria, a beautiful song. It's a spiritual song. She started playing the organ and I started crying uncontrollably. Lorraine was crying, from just seeing me crying, the poor girl. Lorraine saw all the amputations. I had no legs, no right-hand, and I couldn't use my left-hand at all— I couldn't even move my fingers. Sitting in a wheelchair, that was my new reality. I probably cried for about an hour nonstop. It just kept coming out of me. I look back now, and I think that was a normal part of the healing process I had to go through. It also wasn't the first time or the last time. I spent 20 years making a good living and I loved every minute of my job, but it was all gone. I thought I would never golf again. I used to like to go cross country skiing in the mountains and I thought I'd never do that again, either. These were all questions going through my mind—*What am I going to do?* You feel lost, useless, you question yourself, wonder if you would be better off dead. And I wished I was, at that point.

"What are you going to do with me, Lorraine?" I asked. "What am *I* going to do?" I just didn't know.

"We're going to go home," she said, "and we're going to be fine. I will look after you. We are going to be okay." Without Lorraine, I believe I would've just given up, I think I would've committed suicide. And that's no joke.

I was lying in bed at one point, after all of the surgeries were done, and I just didn't have anything to do. I was so bored I just wanted to do *something,* but I couldn't. The doctor came in the room, and he started talking to me.

"Doctor," I said, "I am really bored here. I just hate sitting here not doing anything."

"Would you like to go home?" he asked.

I didn't know what to say, but Lorraine did.

"Let's go!" she said.

She went across the street and bought a pair of sweatpants, a sweatshirt, and she got the wheelchair and we headed to the car, no experience whatsoever with loading me into the car. We had a big, folding wheelchair, but it hadn't been folded in a long time. Lorraine was fighting with it in the parking lot when a security guard and male nurse helped us put it in the trunk. We had a bag of clothing and Lorraine put it on the floor, so I could put my stumps on top. I don't know how she did it! I couldn't help myself at all, but she loaded me in the passenger side of the car, and it was unbelievable. I still remember that I was *so happy* to get out of that hospital. It felt like freedom. I had been there almost seven weeks, so I was so happy just to get out of there, driving down the road. We were coming up on a convenience store and I hadn't smoked in seven weeks, and I wanted a cigarette, *badly*.

"Paul," Lorraine said, "you haven't smoked in over seven weeks! Why don't you quit?"

"Lorraine, I just want a cigarette."

[17] "Do it."

SO LORRAINE got me a pack of cigarettes! Crazy. She had to light it for me, I couldn't even hold it, but it felt *good*—so good, I was dizzy, I loved it, that little freedom. I was just happy to be out there, not being in the hospital anymore. Lorraine was staying at her aunt's home, so that's where we went and spent the night. When we got there, we didn't know what to do. I was in a wheelchair and there was no wheelchair ramp, no provisions like that at all, so three ladies picked me up out of the car and started lifting me up the four steps to the front door! Let me tell you, I was only 140 pounds, but I was so afraid they would drop me. It took them a while, but they got me in the house.

We were sitting around relaxing and I was smoking my cigarettes—having such a blast just smoking my cigarettes. It was unbelievable, just to be outside of the hospital, it felt so good. I didn't keep this up very long, though, because I was always so constipated in the hospital I had taken some prune juice. Well, the smoking and the prune juice and nerves led quickly to Montezuma's revenge[24]. I had diarrhea so bad and we didn't have a bathroom chair, so they had to lift me and put me on the toilet. I must have gone ten times, and it got to where it wasn't funny anymore! Well, funny now, maybe, but not funny back then, definitely not funny for Lorraine and my aunt. They had to lift me onto that toilet every half hour that night.

The doctors didn't give me my meds to go home with, and I was very nervous, panicking, totally out of my zone. Fear set in, big time. I got so sick from it all, it was unbelievable. By the next morning Lorraine was upset, I was upset, and I was crying again. Lorraine and her aunt put me back in the car and we went back to the hospital. It was brutal. They put me back in

---

[24] "Traveler's diarrhea (TD) is a stomach and intestinal infection. . . . It has colloquially been known by a number of names, including Montezuma's revenge and Delhi belly." —Wikipedia

the hospital room. Lorraine left, and I thought she left *me*. She just was tired, though, she needed a break. She was doing a lot for me. I think she had to consider if she was going to go through with all of this. I think it would have only been normal for her to think that through, and I wouldn't have blamed her if she would have left. In fact, I didn't think she was going to stay. But Lorraine stayed. She is still here today, and she is still awesome.

The next day in the hospital, I was taken out of the burn unit. "Paul," the doctor told me, "you're going to Glenrose Rehabilitation Hospital in Alberta tomorrow," and somehow or another I got to the head of the line for rehab. "Since you were not afraid to go home on your own," he said, "—and we were afraid, by the way, that you were going to go home, and we'd never see you again—your next step is to start rehab." They transported me by ambulance to Glenrose the next morning. Lorraine showed up just a little after I arrived there. They brought me into my room and put me in bed, and Lorraine and I settled in. Glenrose was built during the First World War. It was a very old hospital but nice and clean.

"I don't like you being here," Lorraine said. "This place is very old. I just don't want to leave you here."

"It's okay," I said, "it's just the way it is. I have to be here."

The nurses checked up on me. They brought me lunch, and Lorraine fed me. Doctor Trachsel came in to see us and explained that Lorraine could not stay.

"Lorraine, I'm sorry, you are not allowed to come here during the day. The only time you are allowed to visit is after Paul is all finished with his rehab every day, not until four o'clock. He has to learn how to do everything on his own again. He has to learn how to feed himself, wash himself, do his toiletries. He has to learn how to do everything on his own."

Lorraine was in shock. But now that I look back, Dr. Trachsel was a life-saver for me. She did what was right for me. I just disliked her all the way because she was hard on me. She made me do exactly what I had to. There was no self-pity allowed, no feeling sorry for myself. She was hard, and she was tough, and that was hard on me. She explained the staff was there to help me. Lorraine wasn't helping me anymore, and I had to do everything I could myself. If I had to go to the bathroom they would help me because I couldn't clean myself. I needed help eating and they would help me. At 9 a.m. I used to go for physiotherapy for one hour, then at 2 p.m. I had to go for occupational therapy for one hour. Then I was done. After being there for two weeks they said I could go home Wednesday night, come back Thursday, stay Thursday night, go home Friday, Saturday, and Sunday, and come back

Monday. They wanted me to get used to living on my own, and it worked out pretty well. In physiotherapy, we worked on physical skills, relearning how to do things as a disabled person. It was hard, learning how to do things all over again in a very different way. I felt like a 40-year-old child.

I remember Lorraine driving down the road and being afraid of the ice, afraid of running into a ditch, and I felt so useless. I couldn't even get in the car on my own. I didn't even have legs at that point, I had no prostheses yet. They brought me into the prosthetic department and casted me to build my first prostheses. I was somewhat excited because I saw one gentleman in the burn unit who had two artificial hands and two artificial legs. His name was George Uzymirski and he walked really well, I was so impressed. It inspired me so much, it made me believe that I could walk again. It took about a week and I did get my legs. They fitted them on me—it was so different, but I was excited that I was going to walk again. When they were fitting me for the prostheses, I remember telling them, as a joke, that I was six-foot-four inches tall. I wasn't, I was five-foot-eleven before they amputated my feet, but they made my legs that height. The joke was on me! I would be six-foot-four from then on, but it was funny. And I was learning how to laugh again, and laughter heals a lot.

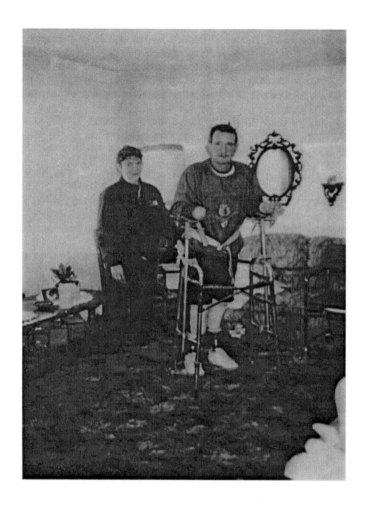

I went home that weekend and started to recover, one step at a time. I would walk ten feet down the hallway, so I had walls to stop me from falling sideways, and I would hang on to Lorraine's shoulders or use a walker, little by little. To tell you the truth, it's like starting all over again. You know what to do, but you can't do it anymore. It was very slow and painful but what choice did I have? I'd have to do the best I could, and that was all I could do. But it was still incredibly frustrating. I used to be able to repair just about anything, for example, but I now didn't have the hands for fixing anything. That's just the way it was. I didn't realize it at the time, and no one explained it to me then, but I was in post-traumatic stress. You have to learn how to manage stress like that, and that's very hard. Fear of the unknown was probably the biggest burden on me. I didn't know at all how my life was going to turn out or what was going to happen day-to-day.

I was anxious to go home and afraid at the same time.

"Maybe I should stay here, with people like me—disabled people," I said to Lorraine.

"Paul," she said, "how are you going to get back to the normal world if you want to live with disabled people?"

"I'm no longer normal!" I said. "I'm a disabled person, no longer capable, no longer able to deal with day-to-day living on my own."

Lorraine thought it over. "I remember," she told me, "Dr. Tredget said you have been through a lot. If you think you'd like to do something or you think you'd like going somewhere, Paul, *do it*. He said because you have been through so much, it's probably taken time off your life. So, if you've wanted to do something in your life, do it. Don't hesitate."

To this day, over 26 years later, I try to live by that philosophy. Sometimes you're limited in what you can do, and you can't only think of yourself. But back then relearning everything was very hard. You try to do something, and you can't. Frustration sets in, and I would panic. I could hardly feed myself. Lorraine helped with everything. Physiotherapy became a regular part of my life, and when not in physiotherapy, people were teaching me how get back into life and how to survive. It wasn't easy, but I did the best I could. My main goal was to go home. I wanted to ski again, and at all costs. Skiing was my passion. When you become disabled what you lose is your passion, and that's what's the hardest.

Meanwhile, we got bad news about my father. He was diagnosed with pancreatic cancer when he was 84, and he got very sick while I was in the hospital, and that was hard. He had also had a stroke when he was 80, making him unable to speak. We hadn't been talking or communicating much. I went to go see him, and we were both upset. Life is unfair, but often all we can do is accept it as it is. We're powerless over other people and situations. But I knew, seeing my father there in the hospital, that I had to love myself, and in that visit, in his own way, my father told me he loved me. When I was young I hated him, I had wished he'd go away, but as I grew older I realized the hardships he had gone through, losing his mom and little brother and being blamed for it. My aunt would later explain the abuse my Dad lived through and it helped me understand him and what he went through himself. Years later, after he had his stroke, he couldn't even speak anymore.

As I walked into the hospital room, he recognized me, and my sister and he started crying. He was trying to speak, and I believe he was trying to say he was sorry. I could see it.

"I'm sorry, Dad," I said. "I wish I could have done better." He was crying a lot. I am crying as I write this.

And I finally got to make peace with my dad.

It's important to make amends, it's morally lifting. I healed my own soul that day. My dad and I had been through hard times, but we made amends that day and it was powerful. My dad loved me, as strange as that might sound.

Hard times have taught me so much.

It was the most important thing I ever did, to go see him and make peace with Dad. I walked away with peace of mind. Sometimes acceptance takes a while.

We lived day by day as I worked to get back to life and my father was in the hospital. The doctors said he would only live for six weeks so I went and saw him as much as I could. It was painful to see him. He was losing ground. They decided to remove his gallbladder, with my brother's approval, but it didn't work in my dad's favor at all. Even though the doctors believed this would extend his life he passed shortly after the operation. I went and saw my father after the surgery. He had aged so much I couldn't recognize him, he looked so old and so tired.

"Is that your father?" Lorraine asked.

"Yes," I said.

"But he looks so tired, so worn out," she said.

It was so sad to see him in that condition, I just couldn't believe it. We left as he was sleeping, and we came back the next day, to find him awake. I walked in the room and he recognized me. I could understand what he was trying to say. He didn't want to go on anymore. He was trying to tell me that he was finished, that he had given up. He had his hands up and he was waving, as if to say, "It's enough, I don't want any more," and I could understand what he was saying. I felt so hopeless. I myself had just been through such sadness. People say you have to learn how to deal such situations but I'm sorry, there's no learning how to deal things like that at all. What I did was grieve. I accepted, I had no choice. These are battles you fight inside yourself, as you hurt and there's nothing you can do about it. People would say to Lorraine, "You must let God help you with this."

"If God is so good," Lorraine would say, angry, "why did he take Paul's feet, his right hand, and most of his left hand?"

And on top of it all I was losing my dad, suffering with cancer. He lived exactly six weeks after he was diagnosed. The doctor asked my brother and me, "If your dad passes away do you want us to revive him?"

"No," we said. "We want him to pass in peace. Dad didn't want to go through this again."

Sad.

I loved my father very much. He was hard on us, but he had a hard life himself. He didn't know any better. My own life has been hard but blessed by the grace of God with great people. I find I can love people who have had a hard time, maybe because I did. I pray for them and it helps me forgive myself, and they hopefully forgive themselves. It's like praying for your enemy—it helps you understand them, and I don't know why but it really works. We all fight our own battles inside, and we have to be nice to other people because we don't know which ones they're fighting, themselves.

[18] SKI LEGS

WE WERE STAYING IN A MOTEL so I could learn how to live on my own, and I was going to the hospital every day for my physiotherapy. I was walking and starting to get along really good, starting to use my hand more, and learning how to drive again. Before my dad passed, I was going to see him at the hospital when I told Lorraine, "I don't think I will ever drive again." She pulled the car over.

"Here," she said, "drive."

I got in the driver seat and I drove for about two miles.

"See?" she said. "Now you can drive."

I was nervous driving with artificial feet, but that was okay, I loved it. It helped a lot to raise my confidence. There I was, going to the hospital each day. I made a decision on my own that I would leave soon, so I told the head nurse about this.

"I'm leaving February 15 to go home, I'm being discharged from the hospital," I said.

"That's wonderful!" she said, "that's great!"

Dr. Trachsel came in and talked to me. "I'm hearing a rumor," she said, "that you're going home February 15? Who came up with that idea?"

I started laughing. "Well, I just thought it would be a good day to go home," I said.

"Yes," she said, "but you haven't done your driving training yet!"

"It's too late," I said.

"What you mean, it's too late?"

"I've been driving," I said.

"No, Paul, you still have to go for driver training. I have to arrange it for you."

I did, and I can't remember exactly, but it was about four days of driver training, and it was really good. I was pretty happy to be driving again. And they discharged me on February 15, 1990.

We had been home about a week when we got the call that my dad wasn't doing well at all. We had seen him the day before, and when we left he was in a coma. When he passed, we had a prayer service for him, because Dad didn't want a funeral. It was very touching. My dad just wanted to be left in peace. I think the thing we desire the most is peace of mind, even as we fight our own battles. To be at peace with yourself is the most important one to wage and win.

The most painful thing was losing the use of my hands, but losing a loved one is terribly painful, too. I will never tell a person he is lucky just because he's fortunate to still be alive. I endured hell on earth. It was very depressing. I was going to stick my chest out and say "I am going to survive! I want to be tougher than this!" Well, that's just not the way it works. Once you've lost your hands, you've lost your life skills, and you lose your identity. I don't know why Lorraine ever stayed. I used to think sometimes that if she would've left and gotten her freedom back, it might've been better for her, but she stayed and that's great. I will be a burden for the rest of my life, that's just the way it is, and it will never change. I will always be disabled. If you're going to be careless and cocky, think again. Do you want to be like me? I'm getting older now and it's getting worse, but that's the way it is.

We were home again, and I was learning to do basic things like showering and going to the bathroom. Lorraine cleaning me after I went to bathroom was the most humiliating thing I've ever lived through. In rehab they asked me, "Is there anything else we can help you with or show you?"

"Yes!" I said. "How do I wipe my butt?"

Because I had only 30 percent left of my left hand, they said they could not help me. It was awful, but what choice did I have? To this day I often use the prayer: *God grant me the serenity to accept the things I cannot change, and the strength to change the things I can. And give me the wisdom to know the difference.*

I was stuck at the "change the things I can" part.

Does the pain go away? Somewhat. I would say you learn how to manage the pain, but it doesn't really go away. You can cry, and I cried a lot. I would go off on my own and just cry, just hurt. The pain would pour out, and that's all you can do. You feel hopeless and helpless.

I tried to go back to work as a supervisor, but the stress was hard on me. I did the best I could. The first summer I was busy, often traveling back to Edmonton for leg adjustments.

We live 280 miles from Edmonton, and I think we put something like 60,000 miles on our car. It was brutal. I didn't know anything about prostheses, either. Sometimes I would just get home after I had gotten an adjustment done in Edmonton and it didn't work. I had to turn around and go back, all the way to Edmonton. We drove, and we drove, it was crazy. And then Workman's Compensation said since I was a ski instructor before the accident, they would build me a set of ski legs. I was so happy!

I had talked to John Gow, who used to be a ski instructor. He lost both of his feet in an airplane crash. Stuck there in the mountains, he froze his feet off. He was a great instructor and he's still a ski instructor. Today I talk to him a lot on the phone. I first met him at Silver Star, British Columbia, and he was a great guy. We didn't get to ski that day, it was too cold, but he was a very encouraging guy and he helped me accept the things that were going on. Nice things, too. You either start accepting or agree to fail, and I didn't want to fail.

They got my legs built and I picked them up just before Christmas. Before my accident I was supposed to work at the ski area and run the ski school, which I didn't do, of course, but as soon as possible, we all went to the ski hill. I strapped on my ski legs and they were very heavy. They took me up on the snowmobile because they didn't want me on the T-bar. I got to the top of the bunny run and I snowplowed all the way down! I was so happy, it felt so good! It was so awesome to ski again, and I was so grateful. It empowered me. It felt so great. Lorraine was there, and everyone was so happy—the whole ski hill was watching and cheering me on, so wonderful to see your friends backing you like that. There were a lot of tears of joy. I think what my friends were really seeing was my recovery, and that made them very happy.

Meanwhile, one of my biggest problems after the accident was sleeping. I just could not sleep, so the doctor referred me to a psychiatrist who explained something about the right and left sides of my brain and depression, ". . . and this is why you cannot sleep," he said. He put me on a type of antidepressant and it did help me. I continued with counseling in those days, but back then they didn't diagnose post-traumatic stress. What we talked about were coping skills. The fact is, as I'm writing this book I'm realizing more now what I've been through, and I'm still going through it. I do have better days, but I do still have depression, something I would say that's just part of my life now, when you get injured at 40 years old.

I was in the prime of my life. I used to ski and teach skiing. I was going for my Level 3 International designation as a skier. I was going to go for my Level 1 Coach designation. I had taken emergency wilderness first aid which is equivalent to an

EMT. He wanted me to work as a winter/Helsinki guide. I had taken my Level 1 avalanche training. I had done a lot of backcountry skiing, mountaineering on Telemark skis. I took a survival course. I was in great physical condition and I loved the outdoors. When I lived in Kenmore some friends wanted me to go rock climbing with them. I had the best life ever. I loved my life and it was over. I had to start all over again. Today, I think I'm doing pretty good considering what I've been through.

That summer was hard, and the hard part was the unknown. But we made it through and when the fall arrived, I went and worked as a ski school manager. I had to start from scratch. We had no instructors, so I visited all the schools around and I made up some ski packages we could sell through them. The packages included all your rental ski equipment with lessons. I delivered them to all the schools around within hundred miles of the ski hill, and that turned out really wonderful. A lot of schools booked with us. We could do up to around 100 kids per day which was awesome for a little ski hill, and it was great business. Over time I brought in ski examiners and ran our own ski instructor courses. We came out with about 10 instructors, three of them full-time and about five part-timers for the weekend.

We used to run slalom races with gates and people had a lot fun. I can't remember how much we charged but it wasn't much—the idea was just to involve people, to have fun. There was a company sponsored by Coke which gave out nice prizes and people loved it. I was getting involved again and I loved it, too. I got to ski quite a bit. I could ski hard pack very well but not as well on powder snow. I had a hard time skiing mogul, so we built our own moguls. We just had great fun. Luke and Albert Johnson became very good friends and I miss these two guys very much today. Elizabeth was a wonderful person. Gilbert Maisonneuve and Claude Ouellette and the young Bourgeois—Roy and Evelyn, and many other young people took part. They were great workers. They were dedicated. They wanted the ski hill to succeed, and it did succeed. It was a great bunch of girls and guys. I was enjoying myself. I was in the ski business again and I loved it.

The following winter after I managed the ski school, and I was approached for the manager's job over the entire ski area. They could see how I organized the ski school and made money there, and I knew I could make money with the ski hill. I got the job and started working really hard, in part because we had to do a lot of snowmaking. The problem with snowmaking is the best time to make snow is when it's coldest at night, when the sun was not shining. You can pump a lot of water and make a lot of snow in those hours. I had worked at Fortress Mountain for Joe Collard. He was a great guy and knew everything about running a ski area. Whenever I had a question I would call Joe and he would tell me what to do.

I made snow in cold as low as 40 below. We made so much snow it was unbelievable. We were pumping 700 gallons a minute at 500-pound pressure and that's a lot of water! You would make enough snow to cover a football field a foot thick in 10 hours—now that's a lot of snow! I would get up at 12 o'clock or two o'clock or three o'clock, it didn't matter. I got up when it was time to make snow. I had a good bunch of guys, and they would be right over helping. We could set up the whole system in probably three quarters of an hour and we always had a snow gun set up in a location where we wanted snow. I was still so stressed from my accident that I worked too many hours that winter, and I had a heart attack, at 42 years old.

PAUL HEBERT

PART FIVE: BACK TO LIFE

[19] SNOWMAKING

I T WAS FEBRUARY 2, 1992. We had just finished making snow. I was running the ski instructors course when a guy twisted his ankle. I still had to take a shift of snowmaking. I was tired, stressed out. I guess I was working hard, trying to get my identity back, but I was overdoing it. Two ski examiners were staying at my house, and I woke up that morning with a chest pain. I took Tylenol. Lorraine made me breakfast, but I wasn't hungry. Elizabeth used to come to my house to ride to the ski hill because she was a ski instructor. She came over that day and we got in my truck and started for the ski hill. I got to the corner about a mile outside of town when the chest pain got really bad and my left arm and my neck were sore. "Elizabeth," I said, "I think I'm having a heart attack. You have to drive me to the hospital."

The poor girl was so nervous! I got out of the driver seat and into the passenger seat and it didn't take her long to get to the hospital. They took me to the emergency room and a nurse I knew was there. I told her I was having a heart attack. She gave me some nitroglycerin and some other pills to settle me down. Lorraine showed up and she was very upset. "Do you think you can work any harder?" she said. "Are you trying to kill yourself?" She was crying. I was crying.

I was afraid. I thought, *This time it's over!* Electrocution is very hard on the heart. They did all the tests on me and it showed that I had had a heart attack. I was released from the hospital three or four days later, stayed home for two or three days, and I went back to the ski hill. It was giving me something to do. I wasn't working too hard, and it didn't take long for the chest pain to go away. I started skiing again and I started looking after myself a little bit better. I even started thinking that I should quit smoking. I cut back. It took me through 1995 to quit smoking, but if you ask me the stupidest thing I've ever done in my life, it's smoking. I had such a hard time quitting it was unbelievable! The pain, the agony, and the fear, it was unbelievable. I was talking to a friend in Peace River about it.

"Paul," he said, "how's the not-smoking going?"

"Well you know, I'm still trying quit," I said.

"If you're only trying you're not quitting," he told me. "You have to quit, and you have to stay quit is what you have to do."

Good advice! I was very addicted to smoking, but how I eventually quit was I had heard someone talk about having to use achievable goals. I started backing down one cigarette a day until I was down to five, four, three, two, then one cigarette per day. I stayed at one a day for a long time and once I got there, I needed that one cigarette at noon and it was everything if you were in my way! When it was time for that cigarette I would've run you over! I used to go buy a pack of cigarettes, smoke one or two, then throw the rest out the window of the car. But three or four hours later I would go find them! Crazy, isn't it? But that's how powerful addictions are. I got a little smarter. I gave the pack of cigarettes to someone I knew, so when I needed a cigarette I would just go bum one off of him or her.

Finally, in 1995, I had my last cigarette. Lorraine wrote me a nice letter that said it didn't matter, that she loved me the way I was and if I couldn't stop smoking, she'd love me anyway. It hit home, and it made me quit. It made me realize I needed a kind of support to actually quit. The best things I have ever done in my life were to quit smoking and quit drinking. I have to say thank you to my higher power for both.

But quitting smoking wasn't over because my lungs reacted. I developed severe allergies. I was so sensitive to everything it was really bad. I had bronchitis so bad I used to have coughing attacks—it was unreal! I would start coughing and I couldn't stop. One time I coughed so much I broke two ribs. Lorraine took me to the hospital, it was so bad. I used to almost pass out, and that's when they started giving me Ventolin (an inhaler). It helps with coughing. Before they got me on it, we had gone to Kenmore, Alberta, and we were going to come back to Columbia through an ice field. There was a car accident on the highway. I was so sick I turned around. We went back to Kenmore and I went straight to the hospital.

"How long since you quit smoking?" the doctor asked.

"Six weeks," I said.

"Well," he said, "if it was just a couple weeks, I would tell you to start smoking a little again because you are going to be very sick. Your lungs are going to react badly because they are not in a good condition."

And I had been sick for a long time. The doctor gave me morphine to control the pain of the broken ribs, given I was still coughing a lot. But I then I walked too far, and I split my left leg open at the bottom, because I couldn't feel it being on the morphine. I spent the next year in a wheelchair trying to heal the split on my stump. I don't know how I didn't end it all right there, because it was so depressing nothing mattered anymore. I just couldn't win. In fact, I was going backwards. It was very painful, mentally. I had no hope. I didn't think I would ever recover. Ultimately, I didn't do it, I didn't end it all, and only because of Lorraine. All I did for a while was sit in my wheelchair and read books. We had two cats, and as I look back now, those cats saved my life. They would play with me. They would lay on me. They were everything to me. They kept me entertained. It's funny, we say things like. "They're just animals," but they're more than that. They helped me survive that wheelchair, for over a year.

From the electrocution I developed severe osteoporosis, which was why I broke two ribs when I was coughing. But I was still fortunate in ways. Dr. Dsouza explained something had to be wrong for me to brake ribs coughing. Sure enough, they diagnosed osteoporosis and put me on a medication called Fosamax. I couldn't walk because my leg was split open, and one way to recover from osteoporosis is to walk and exercise a lot. Since I could do neither I gained a lot of weight. By the time I started walking again I was 235 pounds—and that's too much for me! My bone density was down a lot and I was in bad shape. I just didn't know what to expect anymore. All I wanted was to walk again. They didn't diagnose PTSD back then, but I had it big time. How I lived through all of this is a miracle. I don't know how I survived the wheelchair, I just know the cats sure kept me going because the days were long. I know the deep meaning of boredom, and it's one of the worst things to live through, chronic boredom.

Over the next year I finally healed. I started golfing again, and I was really enjoying myself. I was overweight, but I was walking again. I realized how good it felt to *walk*. I can only imagine, when a person is paralyzed, and he can't walk again— you lose a lot of freedom. And that's what really hurts when you become disabled, you lose your freedom, and you're losing your life skills. As human beings, we like to be independent, we like to survive on our own. I will be dependent on other people for the rest of my life. I say to Lorraine that between my brain skills and her hands we can fix anything together.

After I was in a wheelchair for that period it made us think about a lot of things. We had to get out away from the wintery, icy conditions because I would take bad falls backwards on the ice, even with ice cleats. I would whiplash my neck every time. I fell once and hurt my left arm, and now I have chronic tennis elbow—my only good arm and I was losing it painfully, mentally, every time I fell on the ice. So in the winter of 1995 we took a trip to Arizona. We left on the first of February and we came back March 15. We went to Mesa, Arizona, because Lorraine's mom was there. And we fell in love with Mesa. It's such a beautiful place, we now go back every winter. It's going to be a sad day when we can't go anymore.

We came home to Alberta in 1995 to the coldest winter in history. It was unbelievable. Our plans were to go back to Mesa the following winter, but our youngest daughter wasn't finished with school and I didn't want to leave her home alone. So, the following winter, in 1997, we returned to Arizona with a 27-foot motor home we purchased after my accident. But once we started living in it in Arizona we realized the bed just wasn't big enough, so we found another motor home in Mesa. We found a very cheap 32-foot Rockwood and it was great.

[20] I CAN

I HAVE CONTINUED TO HAVE SURGERIES since the accident, up to recently. I had surgery on my stomach, and as the bones tended to re-grow sideways, they had to re-amputate both legs again. Before the surgery I had done a lot of exercising and I stayed in good shape. I walked as much as I could, I was bench pressing 250 pounds and lifting a lot of weights. I was in very good physical condition and had achieved a good attitude as well. The year prior they had removed a neuroma[25] at end of my left leg, which was very painful to walk on. The surgery was a relief and I was walking again within two weeks. *That was easy!* I thought. I expected the next surgery to be easy as well, but little did I know what was coming.

Coming out of surgery that time was a totally different story. I woke up as they were taking my throat tube out. I thought I was dying. I thought I was choking to death—major panic set in. The nurses were trying to calm me down, but I was so afraid of dying. I was actually okay, but I didn't realize it at the time. And this time there was a lot of pain—*and I mean a lot of pain.* They soon discharged me from the hospital, but when I went home I couldn't walk. Both legs had to heal, and it wasn't easy. I was very emotional, very depressed. I waited two weeks and decided I just had to stand up, had to walk. I did, and it was so painful I had to use a walker to get around. But the pain was unbelievable, it was like starting all over again. All the while I did not realize I had post-traumatic stress, and it set in really bad. But I didn't know what I was going through.

---

[25] A neuroma is a painful condition, also referred to as a "pinched nerve" or a nerve tumor. It is a benign growth of nerve tissue frequently found between the third and fourth toes. It brings on pain, a burning sensation, tingling, or numbness between the toes and in the ball of the foot. The principal symptom associated with a neuroma is pain between the toes while walking.
—http://www.apma.org/Learn/FootHealth.cfm?ItemNumber=987

I didn't find out until the following winter, when I was in Arizona talking to a soldier.

"Do you have a hard time sleeping?" he asked.

"Yes!" I said.

"And how's your memory?" he asked.

"Awful," I said.

He thought about that and asked me, "How about loud noises, do they affect you?"

"It drives me crazy," I said.

"Sounds like post-traumatic stress," he said. "You should look into that."

I did, and I found I had *every* symptom!

The last surgery I had was very hard on me mentally as well. I was really hurting, and I didn't know until that solder enlightened me as to PTSD. And that's when I finally started getting help. One of the worst parts was that I used to be so independent, and I then found myself unable to do anything for myself. I can't tie my shoes. I can't button up my shirt. Loraine has to modify my clothing. Going to the bathroom is better now, but sometimes she has to help me clean myself. Very humiliating. Would you, dear reader, do that for a handicapped spouse? If you would, your marriage can last forever! But if you wouldn't, it's very hard to understand. I know a lot of amputees whose wives left them. And that last surgery had me so depressed I didn't want to live anymore. I thought about ending it all. I couldn't take the mental pain anymore. But the fact that Lorraine never gave up on me, I know, is what kept me going. I started seeing a counselor in Mesa, Arizona, and it has helped a lot.

One of my problems was I had sciatic nerve problems in my lower back and I was in a lot of pain. It hurt *so much*—it went all the way down through my groin, so badly I thought I had pulled my groin. I went to see the doctor and they took x-rays of my lower back and groin. There was a bone spar growing on my hip socket and every time I took a step I could feel it grind in my hip.

"Paul," the doctor said, "I am going give you some pain medication, Percocet, and I want you to take them. Start walking. Go hit golf balls. *Motion is lotion.* Paul, you have to get moving,"

I did and it was *so* painful. And I didn't like the effects of being higher than a kite. I hardly took any Percocet. I went back to see the doctor.

"How are you doing?" he said.

"I am still hurting."

"Did you take the amount I told you to?" he asked.

I said yes.

"Take up to three pills just before you go golfing," he said.

Well, I did. I decided to golf with a league in Mesa, and at first, I planned to golf only eight or nine holes. But once I started moving, by the end of the sixth or seventh hole, I started feeling pretty good. Just the fact that I was back golfing felt pretty good. And since the pain wasn't as bad, I golfed 18 holes. I started going to driving ranges, hitting maybe 30 balls or so just to get going, and I soon started golfing again. But my lower back was still in rough shape. I just couldn't manage that sciatic nerve. We traveled to Hemet, California, and I went to a massage therapist for a deep-tissue massage. My shoulders hurt, my arms hurt, my legs hurt, and I had such bad muscle stiffness. But that's when I started taking massages to control my muscle stiffness. Her name was Tina.

"I have a sciatic nerve in my lower back which is very painful," I told her.

"I can relieve that for you," she said. "Get on your stomach. Now, this is going to hurt. I am going to put direct pressure on your nerves and I will then release them. If it gets to be too much let me know."

It was like a miracle. I felt the release.

"Direct pressure is the only way you can control sciatic nerves," she said.

This was a miracle for me, and I've been getting therapeutic massage ever since.

Many people are depressed and will not admit it. Many don't know what depression is. It's in large part panic—we get angry, we feel fear, and end up in a state of panic. And panic is depression. I didn't know that until I talked to my psychologist—but prior to that a woman once asked me, "Do you have panic attacks?"

"Oh my God," I said, "I *invented* them!"

"Well," she said, "panic attacks are depression."

"Oh my God," I said. And the chain of situations that contributed to my depression all came to me at once: People use anger to empower themselves, our parents showed us how. Use anger enough and the fear will push any kid away or make them very defiant. We all fight our battles—which one are you fighting today?

About two years later I was traveling through Montana and woke up one morning with severe chest pain. Again, to the hospital I went. They checked me out and said it wasn't a heart problem but rather just indigestion, and they gave me a stronger painkiller in syrup form. Many nights I would wake up with indigestion. I was on Prilosec for a long time. Three years later I had chest pains again. I went and saw a heart specialist who put me on blood pressure pills. He said I had a heart murmur (I did have rheumatic fever as a child). With all of these meds I started feeling so hot it was hard to take and I started having coughing attacks. I would almost pass out from coughing. We had returned to Arizona to enjoy the weather in May, but I was so hot and so sick I couldn't go outside.

"Paul," said Lorraine, "if this is making you so sick quit taking these pills!"

So, I did. Three days later I was feeling much better. When we got home to Canada I saw a heart specialist who examined me and took me off of the blood pressure pills. The previous doctor had also put me on beta-blockers because I have too many heartbeats, and I couldn't sleep at night. They gave me sleeping pills, but I then had more chest pains. I was also taking Alka-Seltzer and my other medications to control the heartburn and indigestion. Years later, Doctor Abdul Chinhimi checked me out for sleep apnea, and I was a severe case. The amazing thing is that between the beta-blockers and the sleeping pills I woke up at all! One doctor said "It's a miracle we didn't kill you. You are one tough man to survive what you did." And it *is* a miracle. I was on the *wrong* medications in an attempt to survive over 85 hours of surgeries.

I don't think I'm done here on Earth, not yet. I believe that God has a plan for me and I just keep going. What is important overall is that I'm going through all of this and I still feel good about myself. In fact, I can say I love myself and I love who I am. Not many people can tell themselves that, but I can.

Through my psychologist in my recovery program I have had access to resources that have helped me understand and deal with lots of problems. The biggest problem I have found, at least for me, was to realize *problems do not just go away,* that you have to deal with them. You have to take them on and tackle them. The good news is that's not always as hard as it might seem. I'm proud to say I have accepted and worked on this to the best of my ability. None of us are perfect. We make mistakes.

Another and one of the latest things I have discovered is I now refuse to argue with people. It doesn't do anything, it just creates a bigger argument, and self-control in this world is the biggest thing. I've found that since I stopped fighting with people my resentment level is much lower, much better. If we're resentful, it's because we have fear which quickly turns to anger. And I would say the biggest thing I've ever had to fight in this world was *myself* and my post-traumatic stress. It got really bad—and I mean really bad. Adrenaline levels get very high and you can't sleep. Your memory goes bad and loud noise bothers you. This is what I have to get control of now, what I have to learn how to deal with. Counseling has helped me a lot. No one wants to admit weaknesses, but self-honesty is everything.

I can always talk to my very good friend Jimmy. He's a lifesaver, he's like a brother to me. True friends can be everything as well. Harold Peterson is another true friend—he has really helped me in this world. I have special friends I can go to and talk to about anything at all. I used to talk to Harold every week, he was so inspirational. He died just after my last surgery, when I was going through my post-traumatic stress again, which was really hard on me. Harold was a mentor, like a second father to me—he was everything. I have Jimmy now, and this is what I call the *miracles of life*. I will remember them both forever. And I have had so much vital support from my wife Lorraine, from my family, and from my oldest brother, as well.

But we have to put our pasts behind us. One of the biggest things I ever did was to make amends to my father. I decided to love him just the way he was, and found forgiveness is very powerful, in fact possibly the most powerful thing we can ever do. My plan in writing this book was to get forgiveness and to forgive as well. I wish to communicate with my family again, because I do really love them. Recently I visited my brother and sister-in-law and it was so good just to talk with them.

To forgive and accept each other for what we are and to love yourself—I think these things are everything. They add up to self-acceptance. If you cannot make peace with yourself, you still have work to do. Today, I like what I am doing, but I sometimes do not like what others do to me. This is when I have to learn to say nothing or stay away from certain people. Your own peace of mind is more important, it's your freedom, in fact. It starts with respecting yourself, which is everything.

[21] SUNBREAK

WE STAYED IN OUR ROCKWOOD RV until 2004 when we got a bigger, 38-foot motor home because we were living in it full-time by then. Actually, we have been living in our RV now for 19 years. We enjoy it and we are living in what we can afford—we cannot afford a full-time home *and* living in an RV. We have a good life, though, and we have traveled all over the U.S. by now. One winter we travelled to 12 states. We did all the southern states and it was beautiful. We've been to Florida, Texas, and a lot of the northern states. I am very grateful we got to do this. We have traveled to eastern Canada, New Brunswick, Quebec,

Ontario, Manitoba, Saskatchewan, British Columbia, and all over Alberta. I've never been to Europe, so my goal would be to make one trip there, and to take in all the history.

I started golfing in the Canadian Amputee Golf Tournament at the August Glen Golf and Country Club. Very deep sand trap!

We went to Canmore, Alberta[26] where I had started to golf. I had been golfing for three years and I was just starting to hit

---

[26] Canmore is a town in Alberta, Canada, located approximately 81 kilometres west from Calgary near the southeast boundary of Banff National Park. It is located in the Bow Valley within Alberta's Rockies. — Wikipedia

the ball properly. I was taking lessons and really trying to improve my game. I started breaking the hundreds and golfing in the 90s, so I was hitting the ball well and just starting to really like the game when I had my accident. I had just bought a membership to the Canmore Golf & Country Club. It was a beautiful course and you could see the Rocky Mountains from the course. I loved every minute of it. When we went back to Canmore after my accident in the spring, I had gotten a golf attachment for my right hand. We went to the driving range in Kenmore and Lorraine helped me set up my golf attachment and we got some golf balls. It was so bad—my balance was so off and so weak. I had no core strength left. I only weighed 160 pounds and I could hardly hit the ball at first. I broke down crying, and we gave it up for that day.

Lorraine had a good talk with me and she encouraged me to try another day, and I did a little better. I would go to the driving range in High Prairie and hit as many balls as I could. Amazingly, I started getting better, a little at time. I started hitting the ball almost 150 yards. I was so happy! I started hitting over 180 yards, even 200 yards sometimes. I had to empower myself somehow and this is what kept me going. I never gave up. It was very emotional for me. Sometimes I had a hard time with my self-esteem, but I just kept fighting, fighting to get my identity back. Our good neighbor Paul Labe said to me, "Paul, let's go for a round of golf." We played a small golf course at Lakeside Golf & Country Club. Beautiful course! It was sunny and beautiful, and I started hitting the ball not too bad. We didn't keep score, we just had a lot of fun. I started golfing two or three times a week and loved it.

This is how I got back into the game of golf—one day and one ball at a time. You always have to feel you're improving, because if you're losing it doesn't feel good. It's like playing on a losing team, it doesn't feel good. We always want to go forward, it's like we never want to go backwards, and I had felt a lot of this. I had *failed* really bad—when you make such a mistake in your life it's very hard forgiving yourself. On that job site, I *trusted,* and it cost me a lot. It wasn't easy getting started again but something had to empower me. I had to become a better person. We have to be smarter, yes, and wiser. We have to learn to trust our instincts—as human beings that's sometimes all we have. I have golfed a lot in amputee tournaments in Québec, Ontario, Manitoba, Saskatchewan, British Columbia, Arizona, Minneapolis at the Heseltine (that was a beautiful course). I've golfed a lot of beautiful courses in Alberta, and beautiful courses in Vancouver. I've golfed in Kent Washington, Tennessee, Alabama, California, Washington, Vancouver Island—when I think about it, I have golfed so many wonderful golf courses.

At one time my handicap was a 9.5 (that's great) and I golf at a 12 or 13 handicap now. I just love the game. It keeps me empowered. It's like traveling all over. I love the driving to get there. I love seeing different things. I think the most fun part is meeting different people and the traveling. It's great education. If I could, I would love to go to Australia. I have met wonderful people from Australia and New Zealand, from all over the United States, as well as people from England, Scotland, Ireland, France, the Netherlands, just wonderful people. Golfers are usually nice people, often wonderful people. When we went to Mesquite, Nevada, I won my division with a 239-yard drive. It was great! There were people from all over there—Israel, Australia, all kinds of countries.

Today we travel a lot, my wife and I, with the motor home through Arizona, California, Utah, Montana, Idaho, and other northern U.S. states. We recently stayed in Mesa, Arizona, again in the winter months and just loved it. Lorraine has hiked the Grand Canyon many times, north rim to south rim, and still enjoys it. She's is in great physical condition, too.

One of Lorraine's trips down the Grand Canyon by Havasu Falls. In all she has hiked over 100 miles through the Grand Canyon—very strong lady!

I did do some hiking myself, too. My dream was to hike the Grand Canyon. We got to the top of the canyon together once, but it had snowed, and with artificial legs you cannot walk on ice. You have no feelings in your feet and you can't judge how slippery it is, even with ice cleats. I was afraid of slipping and falling so we put it off. However, I did hike South Mountain in Phoenix through Fat Man's Pass, which was a great hike. I don't remember how many miles it was, but it was (and still is!) an achievement. I loved it. I wish I could've done more hiking.

I was walking a lot, staying in shape. We hiked Saguaro Lake two or three times with friends and it was great. But from walking too much and exercising I developed severe sciatica in my lower back. I have severe muscle stiffness from the electrocution, and I had to learn how to stretch a lot because my body wants to go back into the fetal position as a result. But that's okay, these are all things I have to learn how to deal with. I have to get deep-tissue massages each week, but keeping my flexibility keeps me in life.

In the picture below, we hiked South Mountain in Arizona with my good friends Ann and Peter Feleus. Peter was a great golf buddy, but he passed away at 68 years old, despite being in very good physical condition. I really miss him. Their daughter Karen is a great nurse. If I have medical questions I ask Karen—in fact she teaches other nurses and is just a great lady.

As I came down that steep rock at South Mountain, many people were waiting to help me in case I needed it. They just couldn't believe I was hiking up there. People can be awesome.

It was great, I just loved it.

It was beautiful.

And it reminded me of what that old Indian had told me all those years ago.

That there's a sun above those clouds.

EPILOGUE: PEACE OF MIND

TODAY, I LOOK AT PEOPLE to empower them. I still go to the gym and I work out and I realized, *How many people did I say good morning to?* It was a lot. And I know how much I love communicating with people. If I see someone who is hurting, like this young man I saw the other day, I talk to them. He was coming off of drugs and alcohol, he was really suffering. And I started talking to him and he opened up to me. I explained to him kind of why he was going through it and it made sense to him. I used to have a drinking problem myself, it was a "solution" to all of my problems. I asked him, "You've quit drinking now, right?"

"Yes," he said.

"Good," I said. "And now you know you have to face your problems, you can't put them off with alcohol and drugs anymore. And that's why you're suffering, because you're not drinking to hide your problems anymore. You're facing your problems dead-on."

If you're taking drugs or anything else to make you happy, you're not facing your problems. But if you really do what you're supposed to do, and you do what is right, you're going to *stop yourself from failing*, and you're going to do it yourself. But you have to be honest with yourself first. Sometimes we're weak and temptations are pretty strong, and that in itself is being honest with ourselves. And that's huge.

And I got a smile out of him. He thanked me very much. Just to help people, to put a smile on somebody's face, I think, is pretty important.

I look back at my life, and what I lived through was pretty tough at times, and I think of the people that were good to me, helped me, and supported me. And I have become so grateful for that, there are so many amazing people in my life. The owner of Valard Construction, Victor Budzinski, is such an amazing man. He's just a little guy but I call him a little guy with a big heart. He's always so helpful and so nice to me that I've asked him, "Why have you been so nice to me?"

"Because you're a nice guy," he says.

I've met so many nice people in this world, it empowers me to be grateful for all the people who've helped me. **People have loved and cared for me until I learned to love and care for myself.**

I had pneumonia for a month and I finally called my doctor. "Paul," he said, "I want you to go to the emergency room right away. I want a CT scan of your lungs." They were thinking it was cancer.

I was scared.

But I wasn't scared, really, of dying, I was scared of *losing:* losing my way of life I have going now. I have a great wife, great family, and I don't want to lose that. That's what would really hurt me, to lose all my friends. And it turns out that it was not cancer, and it wasn't pneumonia, it was just that something in my lungs changed. So, I have to go see a specialist, now.

But I'm not afraid of whatever consequences may be about to happen. I think I've lived to about the best of my ability and to my fullest, to have a good life and to get back in things and to not be a—I guess what people would call a "negative." I like to be a positive and try to help as many people as I can, even if just by a smile.

We're all the same, people in this world, and how we treat others, is how they're going to react to us. And if you see somebody that is hurting, if people poke at him, he'll poke back. But what I like is if you're able to help the ones that are suffering the most, then you're really doing something. The happy people, the people who are really successful, need no help. But the ones who are suffering, they need help, and that's really my focus.

I don't really work anymore, but on the job if you bring everybody up to speed and stay away from the negative controls and be positive and compliment people, at least somewhere, and if you have a conversation with people and you don't always try to prove people wrong every time they say something, you're going to get very far. When I talk to people and try to really listen to what they're saying, I learn a lot. They'll tell you what's going on. If you see angry people, what they're telling you is, "I'm really hurting," and that's all. And when I see people that are really hurting, I just take it easy around them and I try to empower them. I want to somehow put a smile on their face, that's all.

I write articles for *Powerline Magazine*. I do speak on occasion for oil companies. I try to talk about how the stronger a team you build, the more successful you're going to be. That's hard to see, sometimes. Greed, temptations, something else will take over and it's lost.

And I think it's true no matter where you go, if it's on the job or anywhere, that by trying to empower people, I've achieved the friendships that I have, and you can, too. I am just so grateful to have so many friends, to have met so many nice people who have helped me through my accident and through my adversities. It's just amazing, I am so grateful. I've come this far because of an attitude of gratitude. I have found real peace of mind, and it is wonderful.

AFTERWORD: TELLING MY STORY

GOLF HAS BECOME a real passion. In 2011, we went to St. George, Utah, to the world games. I golfed okay but not good enough to win anything. But in the long drives one year I hit a 260-yard drive, and it felt pretty good. In Mesquite, Nevada, I hit a drive 239 yards to win the triple-amp division. It's great to compete in these events. It's a good way to empower yourself. I have also golfed with John and Pat Rinehart, a great couple, and we shot a 12-under par to win second place overall in social golf, at the Huntsman World Senior Games. In amputee tournaments, I have won my division many times. I had to back off of the three-day tournaments because of my back—I just can't take that much golfing anymore. A two-day tournament is about all I can do now. But I still enjoy going to the driving range and practicing. Interacting with my friends is very important and I think trying to get back into life is very important.

Long drives in Mesquite, Nevada, 2014

I've been very privileged to meet so many nice people all over Canada and the United States through these travels. I have so many friends it's almost unreal. I've really enjoyed meeting all of them. I was golfing in St. George, Utah, where I met a very nice man named Thomas Burling. We were golfing one day, and he asked me if I was military.

"No," I told him, "I was electrified on hi-voltage."

He looked a little surprised. "You seem to be doing very well," he said.

"Yes," I said, "thank you."

"Have you survived PTSD, then?" he asked me.

"Yes," I said, and I explained that I had gotten help over the years. We kept talking and it was great. He's a brilliant guy. We talked a lot and I found out *he* was the doctor who coined the term, "PTSD!"

When he told me, I seemed to be doing well, it inspired me to start sharing my story, and I have done so now at lots of safety meetings, and with this book. It has helped me deal with it all. And my talks with Thomas are one more part of what I call the *miracles of life*. We stay in touch, and he's become a very important part of my life, become a very good friend. It might not have happened without *getting back into life,* and without golf.

ACKNOWLEDGEMENTS

T
O LIVE THROUGH what I have lived through, it took the blessings of the many people in my life. Life has made me rich in skills for living and in people who have helped. My wife Lorraine is an angel sent to help me. She has always remained by my side. So has Victor Budzinski and his family, my brothers and sisters, my wife's family, the Hachey family, and my power linemen family. Thanks go to Herald Peterson and his wife Viola, and my very good friend Jimmy Suzkiel. Gratitude also to the small town of Falher, Alberta, and the Smoky River municipal district.

Jack Zetller, Sam Either, Wally Ladler, these guys helped me a lot and these are some of the good people that really inspired me along the way. Murray Turner with ACE Construction was another one who inspired me, along with Loel Olsen with North Canada Power Commission.

I am grateful for all the doctors at the McLennan Hospital and for Marilyn Dumont, who very probably saved my life on the flight to Edmonton, when she was so focused she just kept talking to me. She truly listened as I was telling her things I wanted relayed to my wife Lorraine, because I didn't think I was going to live, much less make it to Edmonton and University of Alberta Hospital. The doctors there were simply outstanding—Dr. Larose, Dr. Tredget and his staff—and they too, saved my life.

I am grateful for outstanding friends, for the employees at Valard, who donated money to help us. At the hospital, I received a card that was about two feet high and about a full foot wide which my friends in Canmore and Calgary had signed—I couldn't believe it! It made me realize how many friends I have, which helped.

Thank you, Victor Budzinski. You helped me so much through my accident. You're the best employer I have ever had. I never had to look for and another. You're an unbelievable person, one who really cares about people. Your whole family takes after you like that, too. You truly empower others in hard times. Thank you so much. Victor and family, you're the best family I have ever had.

Thanks to our four kids, Donald, Cindy, Robby, and Courtney. In all the time I've spent in the hospital, they've never gotten into trouble. They were all responsible, awesome kids. They have made our lives easier and I am grateful, thank you. And to Lorraine's mother Mary, thank you. You really stepped up and looked after our family. This was wonderful, thank you so much. Thanks to my sisters, Lucille and Claire, my brothers, Louie and Homer, and my mother, Jacqueline.

Thanks so much to my wife and angel, Lorraine. You carried me when I could not carry myself, and I would not be alive today without you. There have been many times I would have given up, have let go, without you there. You are my life. When I was lost, feared abandonment, when I've felt useless and hopeless, when I felt I'd lost my identity through the amputations, felt I'd lost my own identity as I lost the skills I had worked all my life to attain, you stayed. You helped me recover and empower myself again. You helped me get out of bed, you washed me until I could wash myself again. I don't know what I would do without you. You are my star. I cannot imagine what you have gone through. Lorraine, I am so grateful for what you do for me, you are my angel. I love you.

Lorraine made me thirsty for living again, through her love and with her great strength. She has never given up on me.

As a result, I get to enjoy my beautiful grandkids. My granddaughter Ckyrea is the beautiful little girl walking with me in the picture on the book cover. We are very close, and we have a lot of fun together. And Creedon and Gabriel, our other grandchildren, they are such good kids. I love them all very much and they are in our hearts forever. They are all sun above the clouds to me.

Through the hard times these people were all there for me. They all loved me and carried me until I could love and carry myself again. I am also grateful to live in a country with such a good and caring health system, which greatly helped us through those hard times. In considering how much I put into simply surviving my accident, I sometimes wonder what life might have been like without it. But *life is what we do with what we are presented with, and what we learn from that.* I believe we are all here to be so tested, to survive, to make the best of our situations, and to share what we find with others.

APPENDIX: ARTICLES

LEADERSHIP

I T SEEMS TO ME *encouraging team play* is an effective way of working with young people. And once you're a good team player, you might become a coach yourself. Getting those on a team to help one another, to team people up with different people from time to time, and to help them improve, makes all the difference. I only guide players when they need it, and I don't tell them I'm guiding them, either, I just do it. Keep it as simple as possible. Give positive reinforcement. As a team leader, the most important part of your job is to get to know each player's own capacity, to discover his or her assets on the job.

First tell them what they are doing right, and only then correct what needs to be corrected. Trust me, when they have done something wrong, they know. The worst thing is to focus on it. You bring the worst out of them—anger, even revenge. Some players, when they make a mistake, blame everyone around them. These players don't do well in hard times and they are not good to have around a project. You have to work with them on how to focus their attention in hard times, how to focus on the solution and not the problem. You have to find and focus on the positive. And you do that by observing them—Do they show up ready for work? Are they good communicators? You can tell by the way they focus on job tasks and by what feedback they give you about their jobs.

Give someone a job task and let them figure it out. Help out only when needed. Treat your trades people with respect and give positive feedback. The more you can get them to make decisions on their own, the better they are going to perform, and they will feel good about themselves, too. You're their *coach*. Recognize everyone works differently, and never micro-manage. People never really learn until they decide for themselves. Help them with advice when sought and you will build a team faster than you can imagine because people are very capable of thinking on their own. And all you want is a strong team.

What breaks up good teams is people getting involved in *personalities*. If you're gossiping, what you're doing is breaking up your own team. It's the opposite of keeping focus. When we get too far off track it can become a disaster, even to having fighting or marching off the job, which means of course people are discontent. So how do you keep them focused? You have to stay out of the *personality* and focus on *job tasks only*. If one doesn't fit in pull them aside and have a private talk, but always give them their good points first.

Look for the good team players, but you can even hire those who are not yet good team players, the ones who are hard to deal with. In fact, they are usually the ones who will speak up first in many situations. But even the defiant ones, if you can refocus them soon enough, can become some of your best team players. They might need more attention at first, so hear them out one-on-one. At the same time, do not let them interrupt or take over a situation, either. You go right up to them and tell them, "I want to hear you out." This way you put them on the spot. They tend to embarrass themselves, and they can't interrupt you again since they often don't know what to say.

And once you have them focused, start your job planning. Develop hazard assessments. Working powerline is like driving a car down the road. We never fully eliminate the hazards, we can only control them. Even if we shut the power off we have induction, back feed, and we have traffic problems. That's why we have to be aware of and control the traffic. You often cannot eliminate hazards, so *working safely* is a matter of making people aware of what conditions they have to work in.

And I must say, implementing safety on a site is the hardest thing a person can ever do, because every individual has a different personality, a different fear level. Some people are fearful, and some people are fearless, and the ones that are fearless are the hardest ones to work with. Safety is also the hardest job to understand because not everybody thinks the same way about the hazards. You often have to deal with workers who are very defiant. I call the fearless attitude "bull riding," and I cannot understand why anyone would want to ride a bull. Maybe some people have to *prove* they are not afraid, and this is why they pick dangerous tasks. It's a fault of ego.

Being a good team leader is a full-time job. This why you supervise and why you stay on the job. In fact, as a team leader, I stayed on the job full-time. If I needed tools or equipment, I would send a less experienced worker to get what we needed. I had this skill back then, but I didn't yet have the safety knowledge. I do now.

Good leaders *lead* in hard times, and they keep their teams calm and focused. This is how you can identify good leaders, by observing them in hard times. They keep everyone focused. If you resort to blame or anger, you lose the trust of your workers and you split your team. Everyone worth having wants to do well.

A CULTURE OF SAFETY

SOME PEOPLE DON'T SEEM to understand that we're never bulletproof. It doesn't matter what you're doing, safety is probably the hardest job a person has, but the most important. Yet a lot of people don't seem to have time for safety, they're so busy trying to figure out how they're going to *do* the job. This can all be helped by creating a "safety culture." The worker who is *fearful* is a better thinker than one who is *fearless,* as they don't tend to *think* as much. Players in hockey, for example, who are fearless are usually better on offense—they make good goons. The fearful take fewer risks. Extreme fearlessness can be seen in people who climb mountains. K2 is the highest mountain in the world, and how many have died climbing it? I know it's a real achievement, but the risks are just as real. In line work, we also sometimes see total fearlessness and an addiction to adrenaline. But this is very dangerous on a work site.

There are good places to find the adrenaline rush—in the military, for example. You can join the Air Force, the Navy— for Navy Seals it's a whole different level of excitement, for example, working from heights, working off of helicopters— it's all adrenaline rush. In fact, this is why I became a lineman in the first place. It was a challenge. I remember I got to the top of a 50-foot pole once and it felt *great*. I had done something a lot of people never do, can't do, which is work from great heights. Climbing with lineman spurs took a lot of courage and a lot of balance.

There are many dangerous jobs out there, but I find there are a lot of dangerous people, as well. All we can do is educate them, reorganize their way of thinking. For myself, working as a lineman, the only training we had was in first aid and CPR, and not all employees had this training, either. We were therefore reactive and not proactive, not always emergency ready. And powerline work is one of the most dangerous jobs in the world. Many, to this day, die on the job.

But you can identify and work to eliminate or at least reduce the hazards of the job. For example, in the morning do you sometimes walk around your car and look at the tires to make sure they're okay? Possibly not. Yet it's a very good idea to walk around your vehicle before leaving. Pilots know this very well. Well, I'll tell you something. If I had a chance to go back and change things, all I would change would be my *attitude*. I paid a big price and there's no going back.

Some say you can lead a horse to water, but you can't make him drink, so how do you make one *thirsty*? That's the trick. On job sites, you can compliment your people. Empower them. Make them feel good about themselves. Keep working on changing their thinking. Implement safety in a way that they *want* to do it. Make them plan the job steps, then identify, control, and when possible, eliminate hazards. In most cases, the best we can do is control the hazards, not fully eliminate them. Staying focused on job tasks and staying out of the personalities involved are the best ways you do this. And team work. You have to make people work together, not against one another.

One time I was working with big guy and he said to me, "I don't really like you."

"All we have to do is get our job done right," I said. "We don't have to figure out if we like each other or not."

He thought about that and told me, "I think I like you."

Each day simply ask yourself, "What is my responsibility, here? What's my part to play, and what do I need for today to get my work done—food, clothing? The better I get ready for today the easier I make it on myself. And if I don't look after myself I am setting myself up for failure."

Looking after yourself means controlling your emotional wellness, too. Stay focused on yourself, and if another person tries to get to you through your emotions, don't fall for it, don't react to them. Just recognize some people resort to the old emotional and negative controls. This goes nowhere, of course. They might be laughing as they taunt, but this is not funny at all, just totally out of control. And this when you hear leaders say things like "I am tired of babysitting!" Well, this is exactly when you should make them stop using negative controls and act like adults. Do your own job. Because others act up doesn't mean you have to act up also.

Your biggest asset is people skills. Just because someone is unhappy doesn't mean I become unhappy. People are all different, so keep it simple, easy does it, and most of all, think. The biggest problem I see on work sites is people don't know what is expected of them. As a leader, clarify what is expected of them. Give positive reinforcement. Give direction and advice about your expectations of your job. And as a supervisor reinforce a safe environment at work. Make sure everyone goes home every night to their loved ones.

A culture of safety is everything. Keeping everyone at a good pace is the best practice and part of that culture. On the contrary, rushing or being in a hurry is a *panic* situation. For example, what do you do if you're late for work? We drive a little faster and we get more careless. It's a simple thing but nonetheless very important to start your workday always on time so you don't fall behind, which leads to other problems.

Statistics tell us accidents happen when we repeat job procedures over and over again because our brains get to know job tasks by memory. We stop focusing and put things on "automatic," which is very dangerous. We have to stay focused somehow, and one method is to change job tasks with someone else for better focus. Another is to stay away from angry people if you can because this is the worst situation you can get yourself into, especially in terms of your ability to focus. There is nothing-worse than anger. Anger is a hazard. It distracts everyone, and this leads to disaster. Anger should be identified as a hazard, just as defiance should be. Stay focused.

To implement safety, you explain each job step as needed, you identify job hazards and barriers, and minimize, control, or if you can, eliminate those hazards. An emergency response plan is another large factor. Having one enables you and your team to be proactive rather than reactive. It's like checking your car before a trip. Seventy percent of all accidents are rear-enders. It takes *22 seconds* for a semi to come to a stop, which means if one is 20 feet behind you, he is not likely to be able to stop from hitting you if you have to stop yourself. He's going to be on top of you! We can all do a lot better in terms of safety. Likewise, it takes a car on the average five or six seconds to stop, so allow enough space between you.

In the powerline industry in the early 1900s, *one out of every three* linemen would die. Now I believe it's lower, maybe 30 or 40 people die in the US every year, and while it's a lot better, we can still all do better. If you are only focused on getting successful and making money, I want you to think about how long it took me to learn to deal with my disabilities. I had to focus on *surviving*, to cope with different depressions and all the pain, all the suffering and the surgeries. I *died twice*, once in that ditch and again in the hospital. So how "successful" do you want to be? To what lengths are you willing to go? If you're willing to take chances and kill yourself and maybe others, think again. To this day I struggle.

We have choices to make. I hope you make wise ones, and that you retire a happy person, not having to struggle like I do. Safety is a large part of that. Take your life and safety very seriously. What is your *priority*, wealth or health? It up to you and no one else. The funny part is you can have both, you can work safely and still get a lot done, but you might have to do a better job organizing your day. Planning your work each day is actually very easy. And if you have done everything properly and something happens, it doesn't take long to get back on track because you have a plan in place.

You have to make a commitment to be a safe and responsible person, and you have to ask yourself, "How do I become a safe person?" The answer is always, "I care about my freedom and I want to keep what I have. Further, I do not want to risk the freedom of my loved ones who *might* end up my caregivers! Simply abide by the rules. Start with speed limits, be a good driver, and be courteous, for example. Be proactive and not always reactive. Sometimes safety means just being a nice person. Simple. But it always means caring about your freedom and about other people's freedom as well. If we see something wrong, we fix it. Be a team player. If someone doesn't know safe work practices show them, educate him or her, bring them up to speed. Show them how important freedom is. Without your freedom, you have nothing.

If I would have spent more time and attention on safety and protecting my freedom—my accident only took a *hundredth of a second*—who knows how things might have turned out for me? I do know I would have been very successful because I know I have the smarts to get projects done. I had to become smart to survive. I had to learn how to use a lot of new and different options and skills to survive. It's a never-ending process and it's making a commitment to live the best life we can for ourselves and others. If we just try to please others it doesn't work. You have to do it for yourself at first.

And it's a great feeling to go home each night, safe and healthy, with your reputation intact and a clear conscious. You either do what's right, or you set yourself up to fail. In this regard and others, self-honesty goes a long way.

When I injured myself, what hurt the most was I could not do line work anymore. I lost what I really loved doing and this hurt more than anything. I had lost the freedom to do what I love, in that one way. Once I started climbing my first pole, I was hooked. I remember that first time I climbed a 55-foot pole—I was *on top of the world*. It felt so good—I got that adrenaline rush and I loved it. It went straight into my blood. I think we "line hands" crave that adrenaline rush. It makes us take chances, working with high-voltage all the time, and we have to be careful with that. But it's true that you have to be different to be a lineman.

The wrong kind of fearless makes our jobs very dangerous. All it took in my case was a *twentieth of a second* of electricity and it did enough damage to change my life forever. So I want people to be careful. It's a great job we do but respect it. Don't think you're more powerful than electricity because as I found out, a high-voltage arc is equal to a nuclear blast of *37,000 degrees*, and electricity travels at almost the speed of light. You deserve to treat yourself with respect. Don't lose your, your family's, and your friends' freedom. Please stay safe, please protect your freedom.

I read a book by Clay Brown called *Linehand*, a great book. Two linemen were working in the early morning hours up a pole and one said to the other, "How many stars do you think there are in the sky?"

The other lineman looked up at the still-dark sky, obviously filled with an untold number of stars and said, "What do you mean?"

"This is the Milky Way," said the first lineman. "We have a clear, beautiful evening, with millions—make that billions of stars. In life, if you find what you really love doing it's everything. You have found *your* star." That's a paraphrase, but it really is a great book.

If you found your star as a lineman—or found your star no matter what you do—please don't lose your star! Stay safe. Going home to your families every night is the most important thing. And if you can practice being a lineman—or whatever it is you love to do--safely, you get to keep doing it.

ABOUT THE AUTHOR

". . . his efforts to improve safety and awareness among linemen makes him an important contributor to his field."

—www.linemanmuseum.org/hall-of-fame/paul-hebert/

PAUL HEBERT was born and raised in Alberta, Canada, in an age when you mastered the land you claimed by hand and horses. As a boy, he watched a man climb an electrical pole with his hands and the spurs on his feet, and Paul knew what he wanted to do with his life. Through a tumultuous early life,

Paul survived alcoholism and brawls, and he thrived as a power lineman, enjoying the travel and adventure he found in skiing and the love he found with his new wife, Lorraine.

On September 30, 1989, that all changed. On an emergency call, Paul suffered a jolt of 14,400 volts, which would change his life forever.

As Paul says, "Others carried me until I could carry myself again." He found through the depression and pain as a triple-amputee, there were others that cared, there were important things yet for him to do, and there was "sun above the clouds."

Paul became a sought-after writer of articles and a featured speaker on the topics of safety, leadership, and corporate culture of empowerment. "The ultimate award I've ever received is being inducted into the Power Lineman Hall of Fame for giving my safety presentations and making an impact with my talk, 'Safety is an Attitude.' I spread the message that we respect ourselves, that we look after our well-being, and I ask my audiences, 'What is a more important, wealth or heath?' You have to treat your life as your only investment. You have to protect that investment with everything because without your life you lose all your freedom, and we only go around once."

Today Paul continues to spread his message of personal well-being and safety through articles, appearances, and his book. He golfs, often competitively, and thoroughly enjoys making friends wherever he and Lorraine travel. They have four beautiful children and three grandchildren, and spread their time between Alberta, Canada; Mesa, Arizona; and wherever they decide to go RVing.

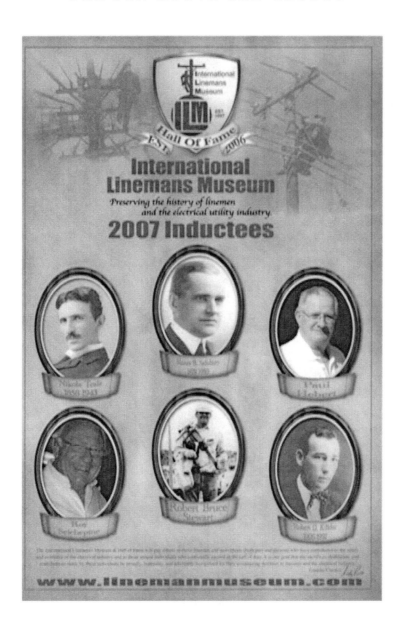

THANK YOU!

I WOULD LIKE to thank all of the following people for their generous support and for making our Publishizer campaign to help launch my book a great success! It means the world to me, thank you!

Alana Ahlskog

Derrick Chudyk

George Harman

Ian Campbell

Jaymee Lucas

Lorie Gill

Michael Drouillard

Sharon Skrepnek

Tammy Burrows

Twila Clay

Vivianne LaPerle

Angele Fontaine

Chris Hite

Chris Wicks

Cynthia Lutz

Edna Owens

Ken Warren

Linda Doucet

Lynn Mabley

Michael Lavis Sr.

Shelden Kirchmayer

Tom MacDuff

Courtney Neufeld

Diana Turcotte

Hubert Lower

Darlene Hachey

David Weakley

Meaghan O'Meara

Rodney Huculiak

Gerald Deason

Bob McNichols

Lee Constantine at Publishizer

Powerlineman Magazine and Byron Dunn

Valard Construction and Victor Budzinski

CPSIA information can be obtained
at www.ICGtesting.com
Printed in the USA
LVOW10s2242110318

569490LV00002B/2/P